Function Theory on Symplectic Manifolds

Volume 34

CRM MONOGRAPH SERIES

Centre de Recherches Mathématiques
Montréal

Function Theory on Symplectic Manifolds

Leonid Polterovich
Daniel Rosen

The Centre de Recherches Mathématiques (CRM) was created in 1968 to promote research in pure and applied mathematics and related disciplines. Among its activities are special theme years, summer schools, workshops, postdoctoral programs, and publishing. The CRM receives funding from the Natural Sciences and Engineering Research Council (Canada), the FRQNT (Québec), the NSF (USA), and its partner universities (Université de Montréal, McGill, UQAM, Concordia, Université Laval, Université de Sherbrooke and University of Ottawa). It is affiliated with the Institut des Sciences Mathématiques (ISM). For more information visit www.crm.math.ca.

American Mathematical Society
Providence, Rhode Island USA

The production of this volume was supported in part by the Fonds de recherche du Québec–Nature et technologies (FRQNT) and the Natural Sciences and Engineering Research Council of Canada (NSERC).

2010 *Mathematics Subject Classification.* Primary 53Dxx; Secondary 57R17, 81S10, 81P15, 22E65, 20F99, 28A10.

For additional information and updates on this book, visit
www.ams.org/bookpages/crmm-34

Library of Congress Cataloging-in-Publication Data

Polterovich, Leonid, 1963– author.
Function theory on symplectic manifolds / Leonid Polterovich, Daniel Rosen.
pages cm. — (CRM monograph series ; 34)
Includes bibliographical references and index.
ISBN 978-1-4704-1693-5 (acid-free paper)
1. Symplectic manifolds. 2. Geometric function theory. 3. Quantum measure theory. I. Rosen, Daniel, 1980- author. II. Title.

QA665.P65 2014
516.3′6—dc23

2014020767

∞ The paper used in this book is acid-free and falls within the guidelines
established to ensure permanence and durability.
Visit the AMS home page at `http://www.ams.org/`

10 9 8 7 6 5 4 3 2 1 19 18 17 16 15 14

Contents

Preface ix

Chapter 1. Three Wonders of Symplectic Geometry 1
1.1. First wonder: C^0-rigidity 1
1.2. Second wonder: Arnold's conjecture 2
1.3. Third wonder: Hofer's metric 6
1.4. The universal cover $\widetilde{\mathrm{Ham}}(M, \omega)$ 13
1.5. More examples: Kähler manifolds 14
1.6. J-holomorphic curves 16
1.7. Marsden–Weinstein reduction 17

Chapter 2. C^0-Rigidity of the Poisson Bracket 21
2.1. Rigidity of the Poisson bracket 21
2.2. Rigidity of symplectomorphisms 29
2.3. Higher Poisson brackets 33

Chapter 3. Quasi-morphisms 35
3.1. Homomorphisms up to a bounded error 35
3.2. Quasi-morphisms and irreversible dynamics 37
3.3. The Poincaré rotation number 39
3.4. The Maslov quasi-morphism 40
3.5. Quasi-morphisms and invariant pseudo-norms 45

Chapter 4. Subadditive Spectral Invariants 49
4.1. The Calabi homomorphism 49
4.2. The action spectrum 50
4.3. Subadditive spectral invariants 53
4.4. Spectral width of a subset 57
4.5. Partial symplectic quasi-states 58
4.6. The Poisson bracket inequality 59
4.7. Two classes of subadditive spectral invariants 61
4.8. Calabi quasi-morphism 62

Chapter 5. Symplectic Quasi-states and Quasi-measures 65
5.1. Symplectic quasi-states 65
5.2. Quasi-states and the quantum-classical correspondence 68
5.3. Topological quasi-states 69
5.4. Quasi-measures 72

5.5. Reduction of symplectic quasi-states 82
5.6. Lie quasi-states 86

Chapter 6. Applications of Partial Symplectic Quasi-states 89
6.1. Symplectic intersections 89
6.2. Lagrangian knots 94
6.3. Applications to Hofer's geometry 97

Chapter 7. A Poisson Bracket Invariant of Quadruples 103
7.1. An invariant of quadruples 103
7.2. Basic properties of pb_4 105
7.3. pb_4 and symplectic quasi-states 105
7.4. A dynamical interpretation of pb_4 107
7.5. pb_4 and deformations of the symplectic form 111

Chapter 8. Symplectic Approximation Theory 119
8.1. The profile function 119
8.2. Behavior at zero 121
8.3. A lower bound via symplectic quasi-states 122
8.4. A lower bound via pb_4 123

Chapter 9. Geometry of Covers and Quantum Noise 127
9.1. Prelude: covers vs. packings in symplectic topology 127
9.2. A Poisson bracket invariant of covers 128
9.3. The Berezin–Toeplitz quantization 129
9.4. Operational quantum mechanics 131
9.5. Classical and quantum registration procedures 141
9.6. Geometry of overlaps and pb_4 143

Chapter 10. Preliminaries from Morse Theory 147
10.1. Spectral numbers of functions 147
10.2. Morse homology 150
10.3. Canonical morphisms 154
10.4. Ring structure 156
10.5. Morse–Novikov homology 158

Chapter 11. An Overview of Floer Theory 161
11.1. Spherically monotone symplectic manifolds 161
11.2. The least action principle 162
11.3. The Floer equation 163
11.4. The dimension of moduli spaces 165
11.5. Compactness breaking mechanism 167
11.6. The Floer complex 169
11.7. Ring structure 171

Chapter 12. Constructing Subadditive Spectral Invariants 173
12.1. Quantum homology 173

12.2. The Frobenius structure 176
12.3. Non-Archimedean geometry of QH 177
12.4. The PSS isomorphism 178
12.5. Spectral invariants in Floer theory 178
12.6. Subadditive spectral invariants revisited 180

Bibliography 185

Notation Index 193

Subject Index 197

Name Index 201

Preface

The symplectic revolution of the 1980s gave rise to the discovery of surprising rigidity phenomena involving symplectic manifolds, their subsets, and their diffeomorphisms. These phenomena have been detected with the help of a variety of novel powerful methods, including Floer theory, a version of Morse theory on the loop spaces of symplectic manifolds. A number of recent advances show that there is yet another manifestation of symplectic rigidity, taking place in function spaces associated to a symplectic manifold. These spaces exhibit unexpected properties and interesting structures, giving rise to an alternative intuition and new tools in symplectic topology, and providing a motivation to study the *function theory on symplectic manifolds*, which forms the subject of the present book.

Recall that a symplectic structure on a $2n$-dimensional manifold M is given by a closed differential 2-form ω which in appropriate local coordinates is given by $\omega = \sum_{j=1}^{n} \mathrm{d}p_j \wedge \mathrm{d}q_j$. The Poisson bracket of a pair of smooth compactly supported functions F, G on M is a canonical operation given by

$$\{F, G\} = \sum_j \left(\frac{\partial F}{\partial q_j}\frac{\partial G}{\partial p_j} - \frac{\partial F}{\partial p_j}\frac{\partial G}{\partial q_j} \right).$$

The Poisson bracket, which is one of our main characters, plays a fundamental role in symplectic geometry and its applications. For instance, it governs Hamiltonian mechanics. The symplectic manifold M serves as the phase space of a mechanical system. The evolution (or Hamiltonian flow) $h_t \colon M \to M$ of the system is determined by its time-dependent energy $H_t \in C^\infty(M)$. Hamilton's famous equation describing the motion of the system is given, in the Heisenberg picture, by $\dot{F}_t = \{F_t, H_t\}$, where $F_t = F \circ h_t$ stands for the time evolution of an observable function F on M under the Hamiltonian flow h_t. The diffeomorphisms h_t coming from all possible energies H_t form a group $\mathrm{Ham}(M, \omega)$, called the group of Hamiltonian diffeomorphisms. For closed simply connected manifolds this group is just the identity component of the symplectomorphism group. The group Ham can be considered as an infinite-dimensional Lie group. The function space $C^\infty(M)$ is, roughly speaking, the Lie algebra of this group, and the Poisson bracket is its Lie bracket.

The structure of the function theory we are going to develop can be illustrated with the help of the following picture. Fix your favorite $t > 0$

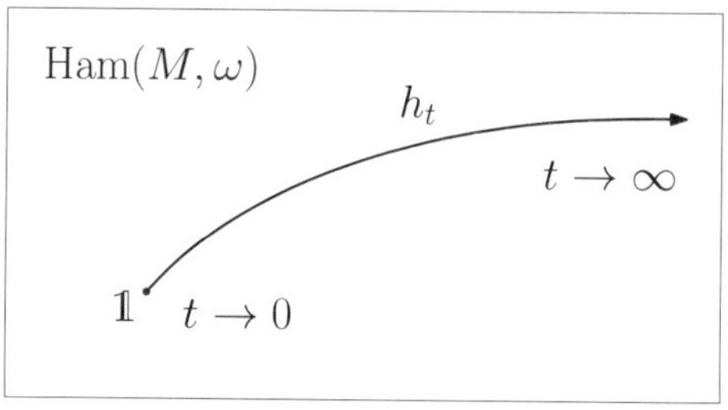

FIGURE 0.1. Two opposite regimes

and consider the natural mapping $C^\infty(M) \to \mathrm{Ham}(M)$ which takes a (time-independent) function H to the time-t map h_t of the corresponding Hamiltonian flow. In principle, this mapping enables one to translate information about Hamiltonian diffeomorphisms (which nowadays is quite a developed subject, see Chapter 4) into the language of function spaces. This naive plan works successfully in two opposite regimes, infinitesimal (when $t \to 0$) and asymptotic (when $t \to \infty$) (see Figure 0.1).

Working in the infinitesimal regime, one arrives at a surprising phenomenon of *C^0-robustness of the Poisson bracket.* Observe that the expression for the Poisson bracket involves the first derivatives of the functions F and G. Nevertheless, the functional $\Phi(F,G) := \|\{F,G\}\|$, where $\|\cdot\|$ stands for the uniform norm of a function, exhibits robustness with respect to C^0-perturbations. In particular, as we shall show in Chapter 2, Φ is lower semi-continuous in the uniform norm. Even though this result sounds analytical in nature, it turns out to be closely related to a remarkable bi-invariant geometry on the group $\mathrm{Ham}(M,\omega)$ discovered by Hofer in 1990. We shall discuss various facets of C^0-robustness of the Poisson bracket. One of them is the *Poisson bracket invariant* of a quadruple of subsets of a symplectic manifold discussed in Chapter 7. Its definition is based on an elementary looking variational problem involving the functional Φ, while its study involves a variety of methods of "hard" symplectic topology. Another facet is symplectic approximation theory, discussed in Chapter 8. Its basic objective is to find an optimal uniform approximation of a given pair of functions by a pair of (almost) Poisson commuting functions.

The asymptotic regime gives rise to the theory of *symplectic quasi-states* presented in Chapter 5. A symplectic quasi-state is a monotone functional $\zeta\colon C^\infty(M) \to \mathbb{R}$ with $\zeta(1) = 1$ which is linear on every Poisson-commutative subalgebra, but not necessarily on the whole function space. The origins of this notion go back to foundations of quantum mechanics and Aarnes' theory of topological quasi-states, an interesting branch of abstract functional analysis. In our context, nonlinear quasi-states on higher-dimensional manifolds are provided by Floer theory, the cornerstone of modern symplectic topology. Interestingly enough, symplectic quasi-states are closely related to *quasi-morphisms* on the group of Hamiltonian diffeomorphisms $\mathrm{Ham}(M,\omega)$.

Roughly speaking, a quasi-morphism on a group is "a homomorphism up to a bounded error." This group-theoretical notion coming from bounded cohomology has been intensively studied in the past decade due to its various applications to geometry and dynamics. We discuss it in Chapter 3. A recent survey of quasi-states and quasi-morphisms in symplectic topology can be found in Entov's ICM-2014 talk [57].

Quasi-states serve as a useful tool for a number of problems in symplectic topology such as symplectic intersections, Hofer's geometry on groups of Hamiltonian diffeomorphisms, and Lagrangian knots. These applications are presented in Chapter 6. In addition, quasi-states provide yet another insight into robustness of the Poisson brackets, see Section 4.6.

Besides applications to some mainstream problems in symplectic topology, function theory on symplectic manifolds opens up a prospect of using "hard" symplectic methods in quantum mechanics. Mathematical quantization and, mostnotably, the quantum-classical correspondence principle provide a tool which enables one to translate basic notions of classical mechanics into quantum language. In general, a meaningful translation of symplectic rigidity phenomena involving subsets and diffeomorphisms faces serious analytical and conceptual difficulties. However, such a translation becomes possible if one shifts the focus from subsets and morphisms of manifolds to function spaces. We present some first steps in this direction in Chapter 9.

The book is a fusion of a research monograph on function theory on symplectic manifolds and an introductory survey of symplectic topology. On the introductory side, the first chapter discusses some basic symplectic constructions and fundamental phenomena, including the Eliashberg–Gromov C^0-rigidity theorem, Arnold's symplectic fixed point conjecture, and Hofer's metric, while in the last three chapters the reader will find an informal crash course on Floer theory. Even though our intention was to make the book as self-contained as possible, the reader is encouraged to consult earlier symplectic literature, such as the classical monographs [107, 108] by McDuff and Salamon. We also refer the reader to the manuscript by Oh [121] on Floer theory. The reader is assumed to have familiarity with basic differential and algebraic topology.

Most of the results presented in the book are based on a number of joint papers by L.P. with Michael Entov. L.P. expresses his gratitude to Michael for long years of pleasant collaboration. Furthermore, some central results of the book are joint with Lev Buhovsky (Poisson bracket invariants and symplectic approximation), Yakov Eliashberg (Lagrangian knots), and Frol Zapolsky (Poisson bracket inequality and rigidity of partitions of unity). L.P. cordially thanks all of them.

Parts of the material have been taught by L.P. in graduate courses at University of Chicago and Tel Aviv University, in a lecture series at UCLA, and (with the assistance of D.R.) in a mini-course at University of Melbourne. We thank these institutions for such an invaluable opportunity. We are indebted to Strom Borman, Lev Buhovsky, François Charette, Adi Dickstein, Michael Entov, Morimichi Kawasaki, Asaf Kislev, Karina Samvelyan, Egor Schelukhin, and Frol Zapolsky for very useful comments on the manuscript, as well as to Dorit Aharonov, Peter Albers, Paul Biran, Paul Busch, Danny Calegari, Yakov Eliashberg, Helmut Hofer, Vincente Humilière, Gil Kalai, Vadim Kaloshin, Yael Karshon, Guy Kindler, Dusa McDuff, Yong-Geun Oh, Yaron Ostrover, Iosif Polterovich, Victor Polterovich, Dietmar Salamon, Felix Schlenk, Paul Seidel, Sobhan Seyfaddini, Ivan Smith, Misha Sodin, Shmuel Weinberger, Amie Wilkinson, and Jinxin Xue for illuminating discussions on various topics of the book. Additionally, we thank the anonymous referees for very helpful critical remarks.

We are indebted to Andrei Iacob for a superb copyediting and very useful critical comments and suggestions on the presentation, and to Carl Sperber for correcting a number of grammatical and stylistic errors.

Finally, special thanks go to François Lalonde, who encouraged us to write this book.

This research has been partially supported by the National Science Foundation grant DMS-1006610, the Israel Science Foundation grants 509/07 and 178/13, as well as the European Research Council Advanced grant 338809.

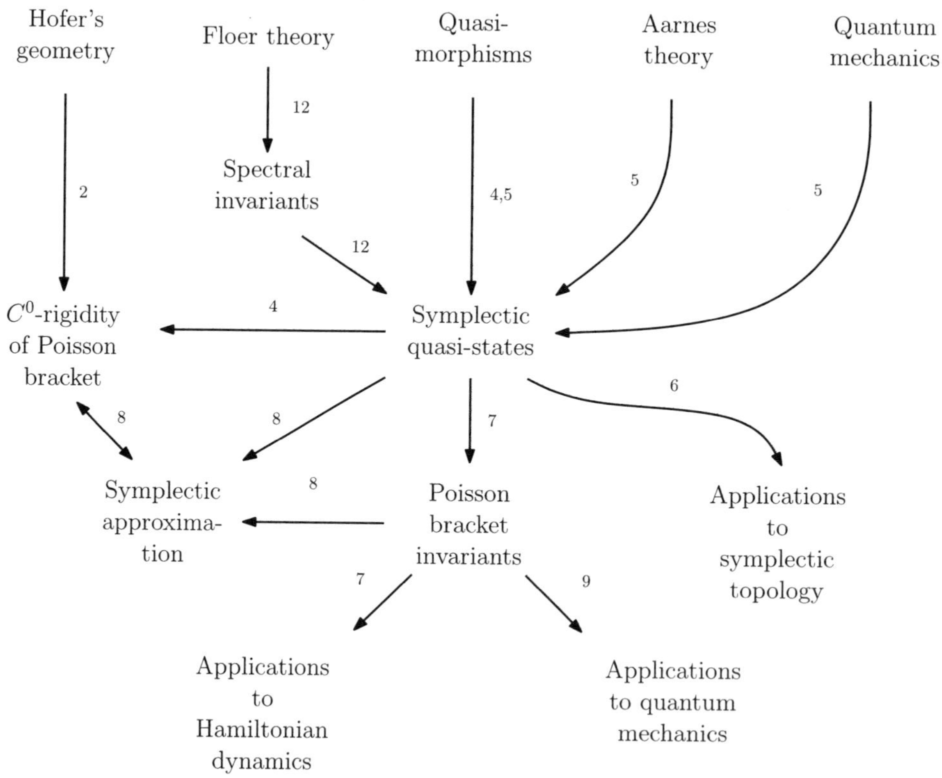

FIGURE 0.2. Subject road map. Numbers next to arrows indicate relevant chapters.

CHAPTER 1

Three Wonders of Symplectic Geometry

This chapter is a mixture of a rapid introduction to symplectic topology and a review of some of its landmark achievements. We mention three wonders of symplectic topology, the first being the Eliashberg–Gromov C^0-rigidity theorem [52, 77], which is the first of several manifestations of symplectic rigidity we will encounter in this book. The second is Arnold's conjecture [4, 6] about symplectic fixed points, which served as a driving force for many modern developments in symplectic topology, among them Floer theory, to which we will return towards the end of the book. The third wonder is Hofer's geometry on the group of Hamiltonian diffeomorphisms [85, 95, 131]. Lastly, we end the chapter with several examples of symplectic manifolds and a brief discussion of J-holomorphic curves.

1.1. First wonder: C^0-rigidity

Let M^{2n} be a smooth connected $2n$-dimensional manifold without boundary, and let ω be a closed 2-form on M which satisfies the following condition:

$$\omega^n = \underbrace{\omega \wedge \cdots \wedge \omega}_{n \text{ times}} \neq 0. \tag{1.1}$$

Then ω is called a *symplectic form*, and (M, ω) a *symplectic manifold.*

Note that ω^n is a top degree form, and hence by (1.1) a volume form, so in particular every symplectic manifold is orientable. The simplest examples include orientable surfaces equipped with area forms and their products. Here the product of two symplectic manifolds (M_1, ω_1) and (M_2, ω_2) is given by $(M_1 \times M_2, \omega_1 \oplus \omega_2)$. More sophisticated examples will be given at the end of this chapter.

A diffeomorphism f of a symplectic manifold (M, ω) is called a *symplectomorphism* if $f^*\omega = \omega$. The symplectomorphisms of (M, ω) form a group with respect to composition. We denote by $\mathrm{Symp}(M, \omega)$ the subgroup of all symplectomorphisms f with compact support: $fx = x$ for all points x outside a compact subset. By the C^k-topology, for $0 \leq k \leq \infty$, on $\mathrm{Symp}(M, \omega)$, and more generally, on the set of all diffeomorphisms of M, we mean the strong Whitney C^k-topology (see [84, Chapter 2]).

Note that symplectomorphisms are defined via their first derivatives, and hence the group $\mathrm{Symp}(M, \omega)$ is by its definition C^1-closed in the group $\mathrm{Diff}(M)$ of all compactly supported diffeomorphisms of M. However, for the same reason, a priori it is not obvious whether it is also C^0-closed.

A natural way to show that a certain class of transformations preserving a given tensor field on a manifold is C^0-closed is to characterize it by the conservation of a "C^0-robust" geometric quantity. Let us illustrate this by the following two examples:

• Let (M, g) be a closed Riemannian manifold. The group $\mathrm{Isom}(M, g)$ of all Riemannian isometries of (M, g) can be characterized by the preservation of the Riemannian distance $d(x, y)$ on M. If a sequence of diffeomorphisms f_k C^0-converges to a diffeomorphism f, we have that $d(f_k x, f_k y) \to d(fx, fy)$ for all $x, y \in M$. Thus if all f_k's are isometries, then $d(fx, fy) = d(x, y)$, and therefore f is also an isometry. We conclude that $\mathrm{Isom}(M, g)$ is C^0-closed in $\mathrm{Diff}(M)$.

• Let (M, σ) be an oriented manifold equipped with a volume form σ. The group $\mathrm{Diff}(M, \sigma)$ of all compactly supported σ-preserving diffeomorphisms of M can be characterized by the preservation of the volume $\int_U \sigma$ of open subsets $U \subset M$. It is easy to conclude from this that $\mathrm{Diff}(M, \sigma)$ is C^0-closed in $\mathrm{Diff}(M)$.

Even though no obvious candidate for such a C^0-robust quantity exists in the case of symplectomorphisms, the above phenomenon persists for symplectic manifolds [52; 77, Section 3.4.4]:

Theorem 1.1.1 (Eliashberg–Gromov rigidity theorem). *Let (M, ω) be a symplectic manifold. Then $\mathrm{Symp}(M, \omega)$ is C^0-closed in the group of all smooth compactly supported diffeomorphisms of M.*

We shall prove this result in Section 2.2 below by using methods of function theory on symplectic manifolds.

1.2. Second wonder: Arnold's conjecture

1.2.1. Mathematical model of classical mechanics. Before studying the properties of symplectic maps, a natural question to ask is whether such maps exist at all. For instance, a generic Riemannian metric on a manifold of dimension ≥ 2 admits no isometries except the identity map (see, e.g., [152, Proposition 1]). It turns out that in symplectic geometry the situation is quite different: an infinite-dimensional group of symplectomorphisms naturally arises within the mathematical model of classical mechanics. To describe this, we first need to discuss some linear algebra.

Let E^{2n} be a real vector space, equipped with an antisymmetric bilinear form $\omega \colon E \times E \to \mathbb{R}$. Define the map

$$I_\omega \colon E \to E^*, \quad \xi \mapsto i_\xi \omega = \omega(\xi, \cdot).$$

Exercise 1.2.1. Prove that the following conditions are equivalent:

(1) $\omega^n \neq 0$.
(2) I_ω is an isomorphism.

If the above equivalent conditions hold, E is called a *symplectic vector space*. For a more detailed account of symplectic vector spaces, we refer the reader to Chapter 4 in [107].

The basic example of a symplectic vector space is $\mathbb{R}^{2n}$, with coordinates $p_1, \ldots, p_n, q_1, \ldots, q_n$, equipped with the symplectic form $\omega_0 = \sum_k \mathrm{d}p_k \wedge \mathrm{d}q_k$, which we sometimes abbreviate to $\mathrm{d}p \wedge \mathrm{d}q$. According to the classical Darboux theorem, it provides a local model for the symplectic structure on an arbitrary symplectic manifold:

Theorem 1.2.2 (Darboux). *Let (M^{2n}, ω) be a symplectic manifold, and let $x \in M$. There exist local coordinates $p_1, \ldots, p_n, q_1, \ldots, q_n$ near x such that in these coordinates we have $\omega = \sum_k \mathrm{d}p_k \wedge \mathrm{d}q_k$.*

Thus, symplectic manifolds have no local invariants, in contrast with Riemannian manifolds, which can be locally distinguished by their curvature tensor. We refer to these local coordinates as *a Darboux chart.* For proofs, see [107, Section 3.1] or [7, Section 43 B].

In classical mechanics, a symplectic manifold (M, ω) plays the role of the phase space of a system. A mechanical system is described by its Hamiltonian, or energy function, $H \colon M \times \mathcal{I} \to \mathbb{R}$. Here $\mathcal{I} \subset \mathbb{R}$ is a time interval which is usually assumed to contain 0. According to a basic principle of classical mechanics, the energy determines the time evolution of the system, in the following way.

We denote $H_t(x) := H(x, t)$. Define[1] the *symplectic gradient*, or Hamiltonian vector field, of H by

$$\operatorname{sgrad} H_t = -I_\omega^{-1}(\mathrm{d}H_t),$$

where $I_\omega \colon TM \to T^*M$ is the bundle isomorphism obtained by applying the isomorphism introduced in Exercise 1.2.1 fiber-wise. That is, for any vector field η on M,

$$\omega(\eta, \operatorname{sgrad} H_t) = \mathrm{d}H_t(\eta).$$

The *Hamilton equation* is the following ordinary differential equation on M:

$$\dot{x}(t) = \operatorname{sgrad} H_t\big(x(t)\big). \tag{1.2}$$

It gives rise to a one-parameter family of diffeomorphisms $\phi_H^t \colon M \to M$, defined by $\phi_H^t(x_0) = x(t)$, where $x(\cdot)$ is the unique solution of (1.2) with initial condition $x(0) = x_0$. (On noncompact manifolds, some extra assumptions on the behavior of H_t at infinity are required in order to guarantee that the solution $x(t)$ exists for all $t \in \mathcal{I}$.) The family $\{\phi_H^t\}$ is called the *Hamiltonian flow* of H, and each diffeomorphism in the family is called a *Hamiltonian diffeomorphism.*

Proposition 1.2.3. *The flow ϕ_H^t preserves ω.*

---

[1] Different authors may use different signs in the definitions of certain notions playing an important role in this book. This includes, in particular, Hamiltonian vector fields and Poisson brackets.

PROOF. The proof is a straightforward computation using Cartan's formula (we use the notation $\mathcal{L}_X$ for the Lie derivative with respect to the vector field X):

$$\frac{\mathrm{d}}{\mathrm{d}t}(\phi_H^t)^*\omega = (\phi_H^t)^*\mathcal{L}_{\mathrm{sgrad}\,H_t}\omega = (\phi_H^t)^*(\mathrm{d}i_{\mathrm{sgrad}\,H_t}\omega + i_{\mathrm{sgrad}\,H_t}\underbrace{\mathrm{d}\omega}_{=0})$$
$$= (\phi_H^t)^*(-\,\mathrm{d}^2 H_t) = 0.$$

Hence, $(\phi_H^t)^*\omega = (\phi_H^0)^*\omega = \omega$ for all $t \in \mathcal{I}$. □

Therefore, any Hamiltonian diffeomorphism is a symplectomorphism of (M, ω).

Example 1.2.4. Consider $\mathbb{R}^{2n}$ with the standard symplectic form $\omega_0 = \mathrm{d}p \wedge \mathrm{d}q$. We interpret q as the position vector, and p as the momentum. In this case the Hamilton equation (1.2) takes the form

$$\begin{cases} \dot{q}_i = \dfrac{\partial H}{\partial p_i}; \\ \dot{p}_i = -\dfrac{\partial H}{\partial q_i}. \end{cases}$$

Consider the particular case of the motion of a particle with mass m in $\mathbb{R}^n(q)$, in the presence of a potential $U(q)$ which depends only on the position. The force in this case is $F = -\partial U/\partial q$. The velocity of the particle is $\dot{q}$ and its momentum is defined by $p = m\dot{q}$. The Hamiltonian of the particle is its total energy $H(p,q) = K(p) + U(q)$, where

$$K = \frac{1}{2}m\dot{q}^2 = \frac{p^2}{2m}$$

is the *kinetic energy*. Thus,

$$H(p,q) = \frac{p^2}{2m} + U(q).$$

Hence, Hamilton's equations are

$$\begin{cases} \dot{q} = \dfrac{p}{m}; \\ \dot{p} = -\dfrac{\partial U}{\partial q} = F. \end{cases}$$

Combining these two equations, we obtain *Newton's second law*: $m\ddot{q} = F$. The passage from solving Newton's equation, a second order ODE in the configuration space $\mathbb{R}^n$, to Hamilton's equation, a first order ODE in the phase space $\mathbb{R}^{2n}$, brings classical mechanics into the framework of the theory of dynamical systems.

One of the first important results in classical mechanics was Liouville's theorem, which states that the time evolution of a mechanical system under Hamilton's equation preserves volume in the phase space. Since the natural volume form on M is, up to a constant factor, ω^n, this theorem follows from Proposition 1.2.3 above.

The discussion above leads us to the following question: *which dynamical features distinguish symplectomorphisms from general volume preserving maps?* One of the answers is given in the next section.

1.2.2. Fixed points of Hamiltonian diffeomorphisms. Let $f\colon M \to M$ be a smooth map. Denote by $\mathrm{Fix}(f)$ its set of fixed points. We say that $x \in \mathrm{Fix}(f)$ is a nondegenerate fixed point if 1 is not an eigenvalue of the differential f_{*x}, or, equivalently, if $\operatorname{graph} f \subset M \times M$ intersects the diagonal $\Delta = \{(y, y) : y \in M\}$ transversally at (x, x). We write $\# X$ for the cardinality of a set X.

Conjecture 1.2.5 (Arnold [4, 6]**).** *Let (M, ω) be a closed symplectic manifold. Let $f\colon M \to M$ be a Hamiltonian diffeomorphism with only nondegenerate fixed points. Then*

$$\# \mathrm{Fix}(f) \geq \dim_{\mathbb{R}} H_*(M; \mathbb{R}) = \sum_k \dim_{\mathbb{R}} H_k(M; \mathbb{R}).$$

Starting from pioneering works by Conley and Zehnder [48] and Floer [67], Arnold's conjecture became a central subject in symplectic topology. It is currently established for all symplectic manifolds, though the amount of subtle technicalities increases drastically when one deals with the general case, and there is no consensus among the experts on whether all of them are completely fixed or not, yet. Fortunately, for a reasonably broad and interesting class of symplectic manifolds the proof of Arnold's conjecture is transparent. In Chapter 11 we will outline an argument based on the powerful machinery of Floer homology.

The link between fixed points of Hamiltonian diffeomorphisms and Morse theory can be clearly seen in the following toy example: Let f_t be the Hamiltonian flow of a *time-independent* Hamiltonian F. Assume further that F is a Morse function, that is, for every critical point of F, $x_0 \in \mathrm{Crit}(F)$, $\det\big(\partial^2 F/\partial x^2(x_0)\big) \neq 0$. Then for any sufficiently small t, f_t has nondegenerate fixed points only. Observe that $\operatorname{sgrad} F(x_0) = 0$ for $x_0 \in \mathrm{Crit}(F)$, and hence $f_t x_0 = x_0$. This means that every critical point of F is a fixed point of f_t. In fact, it follows from a theorem by Yorke [162] that for small values of t the converse statement is true: Every fixed point of f_t is a critical point of F. Thus, by the Morse inequality

$$\# \mathrm{Fix}(f_t) = \# \mathrm{Crit}(F) \geq \dim H_*(M; \mathbb{R}).$$

Remark 1.2.6. Incidentally, Arnold's conjecture illustrates the fact that, in general, not every symplectomorphism is Hamiltonian, even if it is isotopic to the identity through symplectomorphisms. Consider, for instance, the torus $\mathbb{T}^2 = \mathbb{R}^2/\mathbb{Z}^2$ equipped with the standard area form. Here

$$\begin{aligned} \dim H_*(\mathbb{T}^2; \mathbb{R}) &= \dim H_0(\mathbb{T}^2; \mathbb{R}) + \dim H_1(\mathbb{T}^2; \mathbb{R}) + \dim H_2(\mathbb{T}^2; \mathbb{R}) \\ &= 1 + 2 + 1 = 4. \end{aligned}$$

Take a family of (area-preserving!) translations $f_v\colon x \mapsto x+v$, $v \in \mathbb{R}^2$. If $v \neq \mathbb{Z}^2$, the translation f_v has no fixed points, and hence it is not a Hamiltonian diffeomorphism. The last statement can be proved without referring to Arnold's conjecture, see, e.g., [132, Chapter 14] for an elementary proof.

1.3. Third wonder: Hofer's metric

1.3.1. The group of Hamiltonian diffeomorphisms. Let (M, ω) be a symplectic manifold. Having in mind an allusion to classical mechanics (see Section 1.2.1 above), by a Hamiltonian on M we mean a smooth time-dependent function $H\colon M \times \mathcal{I} \to \mathbb{R}$, where $\mathcal{I} \subset \mathbb{R}$ is an interval containing 0. We denote, as before, $H_t(\cdot) = H(\cdot, t)$. A time-independent Hamiltonian is called *autonomous*, in which case it is considered as a function on M. The Hamiltonian flow of H is denoted by φ_H^t or h_t. Observe that after adding to H a function of t we get a new Hamiltonian $H'(x,t) = H(x,t) + a(t)$, for which $\operatorname{sgrad} H' = \operatorname{sgrad} H$, and hence H and H' generate the same dynamics. This ambiguity can be resolved if one deals with normalized Hamiltonians. We call a Hamiltonian $H(x,t)$ *normalized* if

- $\int_M H_t \omega^n = 0$ for all $t \in \mathcal{I}$ in the case when the manifold M is closed;
- $\bigcup_{t \in \mathcal{I}} \operatorname{supp} H_t \subset K$ for some compact $K \subset M$ when M is open.

We denote the space of all normalized *time-independent* Hamiltonians by $\mathcal{H}$.

The maps h_t forming the Hamiltonian flow of a normalized Hamiltonian are called *Hamiltonian diffeomorphisms.* By Proposition 1.2.3, they preserve the symplectic form ω. Denote by $\operatorname{Ham}(M, \omega)$ the set of all Hamiltonian diffeomorphisms. When ω or M are clear from the context, we will sometimes abbreviate this to $\operatorname{Ham}(M)$, or simply Ham.

In classical geometry one often deals with groups of all transformations preserving a given geometric structure on a manifold, such as isometries of a Riemannian metric, volume preserving diffeomorphisms, or symplectomorphisms. Since Hamiltonian diffeomorphisms are not defined via preservation of any structure, it is not immediately clear that they form a group. However, as the name of this section implies, this is indeed the case.

Proposition 1.3.1. *The set* $\operatorname{Ham}(M, \omega)$ *is a group with respect to composition.*

We break the proof of this important proposition into several steps.

Exercise 1.3.2. Let $\phi\colon M \to M$ be a symplectomorphism, and let $F\colon M \to \mathbb{R}$ be a smooth function. Prove that

$$\operatorname{sgrad}(F \circ \phi^{-1}) = \phi_* \operatorname{sgrad} F.$$

Proposition 1.3.3 (Product formula). *Let $\{f_t\}, \{g_t\}$, $0 \leq t \leq 1$, be the Hamiltonian flows generated by normalized Hamiltonians F_t, G_t, respectively. Then the flow $h_t := f_t g_t$ is generated by the normalized Hamiltonian $H_t := F_t + G_t \circ f_t^{-1}$.*

PROOF. We know that

$$\frac{\mathrm{d}}{\mathrm{d}t} f_t x = \operatorname{sgrad} F_t(f_t x), \qquad \frac{\mathrm{d}}{\mathrm{d}t} g_t x = \operatorname{sgrad} G_t(g_t x).$$

Therefore,

$$\begin{aligned}\frac{\mathrm{d}}{\mathrm{d}t} h_t x &= \operatorname{sgrad} F_t(f_t g_t x) + f_{t*} \operatorname{sgrad} G_t(g_t x) \\ &= \operatorname{sgrad} F_t(f_t g_t x) + \operatorname{sgrad}(G_t \circ f_t^{-1})(f_t g_t x) \\ &= \operatorname{sgrad}(F_t + G_t \circ f_t^{-1})(f_t g_t x) = \operatorname{sgrad} H_t(h_t x).\end{aligned}$$ □

Corollary 1.3.4. *Suppose that $\{f_t\}$ is the Hamiltonian flow generated by F_t. Then $\{f_t^{-1}\}$ is the Hamiltonian flow generated by $-F_t \circ f_t$.*

Every Hamiltonian diffeomorphism can be obtained as the time-one map h_1 of the Hamiltonian flow h_t generated by a normalized Hamiltonian $H\colon M \times [0,1] \to \mathbb{R}$. This readily follows from the next proposition.

Proposition 1.3.5 (Change of time in Hamiltonian flows). *Suppose that the Hamiltonian $H_t\colon M \times \mathbb{R} \to \mathbb{R}$ generates the flow $\{h_t\}$. Given any smooth function $a\colon \mathbb{R} \to \mathbb{R}$, the flow $\{h_{a(t)}\}$ is generated by the Hamiltonian $a'(t)H\big(x, a(t)\big)$. In particular, one has the following* time rescaling property: *for a fixed $s \in \mathbb{R}$, the flow $\{h_{st}\}$ is generated by the Hamiltonian sH_{st}.*

The proof is left as an exercise. Using Proposition 1.3.3 and Corollary 1.3.4 we can now prove that $\operatorname{Ham}(M, \omega)$ is a group.

PROOF OF PROPOSITION 1.3.1. Indeed, if $f, g \in \operatorname{Ham}(M,\omega)$ are the time-one maps of the normalized Hamiltonians F, G, respectively, then by Proposition 1.3.3, fg is the time-one map of the normalized Hamiltonian $F_t + G_t \circ f_t^{-1}$. Additionally, if f is the time-one map of F, then f^{-1} is the time-one map of the normalized Hamiltonian $-F_t \circ f_t$. Therefore, $\operatorname{Ham}(M,\omega) \subset \operatorname{Symp}(M,\omega)$ is a subgroup. □

By definition, if $f \in \operatorname{Ham}(M,\omega)$, there exists a path $\{f_t\} \subset \operatorname{Ham}(M,\omega)$ such that $f_0 = \mathbb{1}$ and $f_1 = f$. Consider now an arbitrary path $\{f_t\} \subset \operatorname{Ham}(M,\omega)$ which starts at $\mathbb{1}$. Although each f_t is a Hamiltonian diffeomorphism, a priori there is no reason why $\{f_t\}$ itself should be the Hamiltonian flow of some Hamiltonian. However, this turns out to be true, due to a theorem by Banyaga [13, Proposition II.3.3; 107, Proposition 10.17].

Theorem 1.3.6. *Every smooth path in* $\operatorname{Ham}(M,\omega)$ *based at the identity is the Hamiltonian flow generated by some normalized Hamiltonian.*

An elementary particular case of this statement is valid in the following more general context: Let (M,ω) be a symplectic manifold (open or closed). Recall that $\operatorname{Symp}(M,\omega)$ stands for the group of all compactly supported symplectomorphisms of (M,ω). Denote by $\operatorname{Symp}_0(M,\omega)$ its identity component. Since $\operatorname{Ham}(M,\omega)$ is by definition path connected, we have the inclusions

$$\operatorname{Ham}(M,\omega) \subset \operatorname{Symp}_0(M,\omega) \subset \operatorname{Symp}(M,\omega).$$

In general, $\operatorname{Ham}(M,\omega) \neq \operatorname{Symp}_0(M,\omega)$. However, in some cases these two groups coincide:

Proposition 1.3.7. *Suppose that the first cohomology with compact support of M vanishes, that is $H^1_c(M;\mathbb{R})=0$. Then* $\operatorname{Ham}(M,\omega)=\operatorname{Symp}_0(M,\omega)$.

PROOF. Let $f\in\operatorname{Symp}_0(M,\omega)$. Then there is a path $\{f_t\}\subset\operatorname{Symp}_0(M,\omega)$ such that $f_0 = \mathbb{1}$ and $f_1 = f$. Define the time-dependent vector field ξ_t by

$$\xi_t(f_t x) = \frac{\mathrm{d}}{\mathrm{d}t} f_t x.$$

Then

$$0 = \frac{\mathrm{d}}{\mathrm{d}t} f_t^*\omega = f_t^*\mathcal{L}_{\xi_t}\omega.$$

Since $\mathrm{d}\omega = 0$, by Cartan's formula we obtain $\mathrm{d}i_{\xi_t}\omega = \mathcal{L}_{\xi_t}\omega = 0$, that is, for every t the form $i_{\xi_t}\omega$ is closed. Since $H^1_c(M;\mathbb{R}) = 0$, these forms are exact: $i_{\xi_t}\omega = -\mathrm{d}F_t$, where $F_t\colon M \to \mathbb{R}$ has compact support. Therefore $\xi_t = \operatorname{sgrad} F_t$, and the path $\{f_t\}$ is the Hamiltonian flow of the time-dependent Hamiltonian F_t. □

Although in general the inclusion $\operatorname{Ham} \subset \operatorname{Symp}_0$ is strict (see Remark 1.2.6 above), the difference between the two groups is 'not too big'. The next statement, which was an important open problem known as the flux conjecture, was finally settled by Ono [122]:

Theorem 1.3.8. *For every closed symplectic manifold (M,ω)*

$$\operatorname{Symp}_0(M,\omega)/\operatorname{Ham}(M,\omega) = H^1(M;\mathbb{R})/\Gamma,$$

where $\Gamma \subset H^1(M;\mathbb{R})$ is a discrete subgroup.

For instance, for the torus $(\mathbb{T}^2, \mathrm{d}p\wedge\mathrm{d}q)$ the group Γ is simply the integral lattice $H^1(\mathbb{T}^2;\mathbb{Z}) \subset H^1(\mathbb{T}^2;\mathbb{R})$, see, e.g., [132].

Exercise 1.3.9. The group $\operatorname{Ham}(M,\omega)$ acts on M by its definition. Prove that when M is connected this action is transitive. *Hint*: First prove the result for the open ball $B^{2n}(r) \subset \mathbb{R}^{2n}$. For the general case, use the previous one and Darboux's Theorem 1.2.2 to prove that for any $x \in M$, the orbit $\{f(x) : f \in \operatorname{Ham}(M,\omega)\}$ is open. Deduce the result.

1.3.2. Ham as a Lie group. A useful viewpoint to adopt is to consider $\mathrm{Ham}(M, \omega)$ as an infinite-dimensional Lie group. Its Lie algebra, which is by definition the tangent space at the identity $T_{\mathbb{1}}\,\mathrm{Ham}$, consists of all (time-independent) Hamiltonian vector fields, and therefore can be identified with the space $\mathcal{H}$ of all normalized autonomous Hamiltonians on M. Further, we identify the tangent space to Ham at an arbitrary point g with the Lie algebra via right translations:

$$T_g\,\mathrm{Ham} = (R_g)_* T_{\mathbb{1}}\,\mathrm{Ham},$$

where

$$R_g \colon f \mapsto fg.$$

For example, consider a path $\{f_t\}$ generated by a normalized Hamiltonian F_t. Let us express the tangent vector

$$\xi := \left.\frac{\mathrm{d}}{\mathrm{d}t}\right|_{t=s} f_t \in T_{f_s}\,\mathrm{Ham}$$

in terms of the Hamiltonian. By our identification, ξ corresponds to

$$\left(R_{f_s}^{-1}\right)_* \xi = \left.\frac{\mathrm{d}}{\mathrm{d}t}\right|_{t=s} R_{f_s}^{-1} f_t = \left.\frac{\mathrm{d}}{\mathrm{d}t}\right|_{t=s} f_t f_s^{-1} = \mathrm{sgrad}\, F_t \circ (f_t f_s^{-1})|_{t=s} = \mathrm{sgrad}\, F_s.$$

That is, $(\mathrm{d}/\,\mathrm{d}t)|_{t=s} f_t$ corresponds to $F_s \in \mathcal{H}$. Note that here s is fixed, and F_s is thought of as an autonomous Hamiltonian.

Next, let us find the adjoint action of Ham on its Lie algebra. Let $f \in \mathrm{Ham}$, $G \in \mathcal{H}$ and let $\{g_t\}$ be the Hamiltonian flow of G. By definition,

$$\mathrm{Ad}_f(G) = \left.\frac{\mathrm{d}}{\mathrm{d}t}\right|_{t=0} f g_t f^{-1} = \mathrm{sgrad}(G \circ f^{-1})$$

by Exercise 1.3.2. By a slight abuse of notation we write $\mathrm{Ad}_f(G) = G \circ f^{-1}$.

Finally, let us calculate the Lie bracket on the Lie algebra $\mathcal{H}$. Take $F, G \in \mathcal{H}$, and let $\{f_t\}$ be the Hamiltonian flow of F. Again, by definition,

$$\begin{aligned}\{F, G\} = \left.\frac{\mathrm{d}}{\mathrm{d}t}\right|_{t=0} \mathrm{Ad}_{f_t} G = \left.\frac{\mathrm{d}}{\mathrm{d}t}\right|_{t=0} G \circ f_t^{-1} &= -\,\mathrm{d}G(\mathrm{sgrad}\, F) \\ &= \omega(\mathrm{sgrad}\, G, \mathrm{sgrad}\, F).\end{aligned}$$

This Lie bracket is called the *Poisson bracket* on $\mathcal{H}$. It can be rewritten as

$$\{F, G\} = \mathrm{d}F(\mathrm{sgrad}\, G).$$

The Poisson bracket was defined above for normalized Hamiltonians $F, G \in \mathcal{H}$, but the definition can be extended to any smooth functions $F, G \in C^\infty(M)$ by the same formula:

$$\{F, G\} = \omega(\mathrm{sgrad}\, G, \mathrm{sgrad}\, F) = \mathrm{d}F(\mathrm{sgrad}\, G).$$

Exercise 1.3.10. Show that in $(\mathbb{R}^{2n}, \mathrm{d}p \wedge \mathrm{d}q)$, the Poisson bracket takes the form

$$\{F, G\} = \sum_j \left(\frac{\partial F}{\partial q_j} \frac{\partial G}{\partial p_j} - \frac{\partial F}{\partial p_j} \frac{\partial G}{\partial q_j} \right).$$

Two functions $F, G \in C^\infty(M)$ are said to be *Poisson commuting* if their Poisson bracket vanishes identically, $\{F, G\} \equiv 0$. Note that if F and G Poisson commute, then F is constant along trajectories of the Hamiltonian flow of G (and vice versa).

Exercise 1.3.11. Let F, G be Poisson commuting functions, and denote their Hamiltonian flows by f_t and g_t, respectively. Prove that the flows commute:

$$f_t g_s = g_s f_t, \quad \forall s, t.$$

Prove that the flow $f_t g_t$ is generated by the Hamiltonian $F + G$.

Exercise 1.3.12. Prove that for any $F, G \in C^\infty(M)$

$$\mathrm{d}F \wedge \mathrm{d}G \wedge \omega^{n-1} = -\frac{1}{n}\{F, G\}\omega^n.$$

Deduce that the Poisson bracket of any two Hamiltonian functions has zero mean.

Exercise 1.3.13. Consider the unit sphere

$$S^2 = \{x_1^2 + x_2^2 + x_3^2 = 1\} \subset \mathbb{R}^3$$

with the induced Euclidean area form. Use Exercise 1.3.12 to show that the Poisson brackets of the coordinate functions x_1, x_2, x_3 are given by the relation $\{x_1, x_2\} = -x_3$ and its cyclic permutations.

Exercise 1.3.14. Let (M, ω) and (N, σ) be two symplectic manifolds. Prove that a diffeomorphism $f\colon M \to N$ is a symplectomorphism if and only if it preserves the Poisson bracket, that is, for any $\phi, \psi \in C^\infty(N)$ one has

$$\{f^*\phi, f^*\psi\} = f^*\{\phi, \psi\}.$$

1.3.3. Hofer's metric on $\mathrm{Ham}(M, \omega)$. Recall that a *Finsler metric* on a manifold is a smooth assignment of a norm on each tangent space. Let G be a Lie group with Lie algebra $\mathfrak{g}$. Suppose that $\mathfrak{g}$ is equipped with a norm $\|\cdot\|$ which is invariant under the adjoint action. Then we can extend $\|\cdot\|$ to T_gG for all $g \in G$, by right or left translations, and by the invariance of $\|\cdot\|$ both yield the same result. In this way G becomes a Finsler manifold, which enables us to introduce in the standard way various geometric notions on G. First, define the length of a curve:

$$\mathrm{length}\{g_t\} = \int_0^1 \|\dot{g}_t\|\, \mathrm{d}t.$$

Further, for $\phi, \psi \in G$ put

$$d(\phi, \psi) = \inf\{\mathrm{length}\{f_t\} : f_0 = \phi, f_1 = \psi\}.$$

The function d on $G \times G$ satisfies the axioms of a pseudo-metric, namely

- $d(\phi, \psi) \geq 0$;
- $d(\phi, \psi) = d(\psi, \phi)$;
- $d(\phi, \psi) + d(\psi, \theta) \geq d(\phi, \theta)$;

and the additional axiom of *bi-invariance*:

- $d(\theta\phi, \theta\psi) = d(\phi\theta, \psi\theta) = d(\phi, \psi)$.

Note that the only property d is a priori lacking in order to be a genuine metric is nondegeneracy:

- $d(\phi, \psi) > 0$ for all $\phi \neq \psi$.

Let us return to the group $\mathrm{Ham}(M, \omega)$. Here the Lie algebra is $\mathcal{H}$, and we have several choices of invariant norms on it, such as the L_p norms for $1 \le p < \infty$ or the uniform norm $\|F\| = \max|F|$.

Theorem 1.3.15. *For the uniform norm, the pseudo-metric d is a genuine metric.*

This metric was introduced in 1990 by Hofer (see [85]), who proved its nondegeneracy for the linear symplectic space $\mathbb{R}^{2n}$. Later on, this result was extended to some more complicated symplectic manifolds in [131], and finally proved in full generality in [95]. All the proofs involves methods of "hard" symplectic topology. We present a proof for closed symplectic manifolds in Chapter 4 (see Corollary 4.7.2). We shall refer to the metric d associated to the uniform norm as the Hofer metric. Throughout the book we reserve the notation $\|\cdot\|$ for the uniform norm and d for the Hofer metric. Remarkably, Hofer's metric is essentially the unique nondegenerate bi-invariant Finsler metric on the group Ham [35]:

Theorem 1.3.16 (Buhovsky–Ostrover). *Let (M, ω) be a closed symplectic manifold. Let $\|\cdot\|'$ be a* Ham*-invariant norm on $\mathcal{H}$ which is continuous in the C^∞-topology, and let d' be the associated bi-invariant pseudo-metric on* Ham*. Then d' is nondegenerate if and only if the norm $\|\cdot\|'$ is equivalent to the uniform norm: there exist constants $C, c > 0$ so that*

$$c\|\cdot\| \le \|\cdot\|' \le C\|\cdot\|.$$

The existence of a bi-invariant Finsler metric on $\mathrm{Ham}(M, \omega)$ becomes especially surprising when confronted with the following two results. The first is a fundamental algebraic property of the group of Hamiltonian diffeomorphisms [13]. Recall that a group is called *simple* if it contains no nontrivial normal subgroups.

Theorem 1.3.17 (Banyaga). *Let (M, ω) be a closed symplectic manifold. Then* $\mathrm{Ham}(M, \omega)$ *is a simple group.*

The second result is a general fact about bi-invariant Finsler metrics on simple *finite-dimensional* Lie groups:

Proposition 1.3.18. *Let G be a simple finite-dimensional Lie group with a bi-invariant Finsler metric. Then G is compact.*

PROOF. It is a classical fact [109, Corollary 21.5] that if G admits a bi-invariant *Riemannian* metric and its Lie algebra $\mathfrak{g}$ has trivial center, then its Ricci curvature is strictly positive and bounded away from zero, and hence the group is necessarily compact. Let us construct a bi-invariant Riemannian metric on G. Consider the unit ball $K \subset \mathfrak{g}$ with respect to the given Finsler metric. Let E be the unique ellipsoid of maximal volume contained in K (E is called the John ellipsoid, see [16, Section V.2]). Then E is the unit ball of a Euclidean norm on $\mathfrak{g}$. Moreover, by the uniqueness of the John ellipsoid, E is invariant under the adjoint action of G on $\mathfrak{g}$, and hence defines a bi-invariant Riemannian metric on G.

It remains to show that the Lie algebra $\mathfrak{g}$ has trivial center. Indeed, denote its center by $\mathcal{Z}$, and let $H = \exp \mathcal{Z}$ be its image under the exponential map. We note that for any $X \in \mathcal{Z}$ and $Y \in \mathfrak{g}$, X and Y commute, and hence

$$\exp(X)\exp(Y) = \exp(X+Y) = \exp(Y)\exp(X). \tag{1.3}$$

Therefore, if we let $U \subset G$ be a neighborhood of the identity which lies in the image of the exponential map, then (1.3) implies that $HU = UH$, or in other words, U normalizes H. But since G is connected, it is generated by U, and hence H is a normal subgroup of G. Since G is simple, H is either trivial or all of G. But the latter cannot hold, since it would imply that G is abelian, again contradicting its simplicity. Therefore, H is trivial, and hence $\mathcal{Z}$ is trivial as well. This completes the proof. □

The existence of Hofer's metric on Ham is thus a purely infinite-dimensional phenomenon. Let us mention also that for a compact manifold M, every degenerate pseudo-metric on $\mathrm{Ham}(M,\omega)$ vanishes identically. This follows from Theorem 1.3.17 and the following elementary observation:

Proposition 1.3.19. *Let G be a simple group, and let d be a bi-invariant pseudo-metric on G. If there exists $g \in G$ such that $d(\mathbb{1}, g) > 0$, then d is a metric.*

PROOF. Define

$$K = \{h \in G : d(\mathbb{1}, h) = 0\}.$$

Using the triangle inequality, K is easily seen to be a subgroup of G, and by the bi-invariance of d, it is a normal subgroup. By the simplicity of G, either $K = G$ or $K = \mathbb{1}$, and since $g \notin K$, the latter holds. That is, for any $h \neq \mathbb{1}$, $d(\mathbb{1}, h) > 0$. □

1.3.4. A small scale in symplectic topology. There is a natural 'small scale' for sets in symplectic topology. The following notion is motivated by [85].

Definition 1.3.20. A set $X \subset M$ is called *displaceable* if there exists a Hamiltonian diffeomorphism $\varphi \in \mathrm{Ham}(M,\omega)$ such that

$$\varphi(X) \cap \mathrm{Closure}(X) = \varnothing.$$

Example 1.3.21. Let us illustrate this notion in the following basic example. Consider the sphere S^2 equipped with an area form ω of total area 1. A simple closed smooth curve on S^2 is called *an equator* if it divides the sphere into two discs of equal areas.

Exercise 1.3.22. Show that every equator of S^2 is nondisplaceable.

Exercise 1.3.23. Show that the group $\mathrm{Ham}(S^2)$ acts transitively on the space of equators. *Hint*: Let L_0, L_1 be two equators. Choose an orientation preserving diffeomorphism ϕ with $\phi(L_0) = L_1$. Modify if necessary ϕ so that $\phi^*\omega$ coincides with ω on $T_{L_0}S^2$. A version of Moser's argument [107, Section 3.2; 114] yields a diffeomorphism ψ of S^2 so that ψ point-wise preserves L_0 and $\psi^*\phi^*\omega = \omega$. Thus, $\phi\psi$ is a symplectomorphism of S^2 taking L_0 to L_1. It remains to note that the group of symplectomorphisms $\mathrm{Symp}(S^2)$ of S^2 coincides with $\mathrm{Ham}(S^2)$. Indeed, Moser's argument shows that the group $\mathrm{Diff}_+(S^2)$ of orientation preserving diffeomorphisms of S^2 acts transitively on the space of symplectic forms on S^2 which are compatible with the orientation and have total area 1. The latter space is contractible (why?), and hence $\mathrm{Diff}_+(S^2)$ retracts to the stabilizer of the standard symplectic form, that is, to $\mathrm{Symp}(S^2)$. But the group $\mathrm{Diff}_+(S^2)$ is known to be path connected, see Exercise 1.5.3 below. Thus $\mathrm{Symp}(S^2) = \mathrm{Symp}_0(S^2)$. Finally, apply Proposition 1.3.7 to conclude that $\mathrm{Symp}_0(S^2) = \mathrm{Ham}(S^2)$.

On the other hand, let $B \subset S^2$ be a closed disc with smooth boundary of area $< \frac{1}{2}$. We claim that B is displaceable. Indeed, fix the standard model of the sphere,

$$S^2 = \{x_1^2 + x_2^2 + x_3^2 = 1\} \subset \mathbb{R}^3, \quad \omega = \frac{1}{4\pi}\Omega,$$

where Ω is the standard spherical area form. It readily follows from Exercise 1.3.23 that there exists $\phi \in \mathrm{Ham}(S^2)$ such that $\phi(B)$ lies in the upper hemisphere $\{x_3 > 0\}$ (prove this!). The latter is displaceable by the rotation by 180 degrees around the x_1-axis. Therefore, the disc B is displaceable.

Example 1.3.24. The product of equators in $(S^2 \times S^2, \omega \oplus \omega)$ is not displaceable. This can be shown with the help of Lagrangian Floer homology (see [117]). In Section 6.1.1 we extend this result to certain singular subsets of $S^2 \times S^2$ and prove it by using tools of function theory on symplectic manifolds.

Following Hofer, we define an important invariant of a displaceable subset $U \subset M$, *Hofer's displacement energy* $e_H(U)$, as $\inf d(\mathbb{1}, h)$, where the infimum is taken over all $h \in \mathrm{Ham}(M, \omega)$ displacing U. Remarkably, Hofer's displacement energy of every open displaceable subset is strictly positive. We shall prove this for closed symplectic manifolds in Chapter 4 (see Corollary 4.7.2). We follow the convention that $e_H(U) = +\infty$ if U is nondisplaceable.

1.4. The universal cover $\widetilde{\mathrm{Ham}}(M, \omega)$

In the present book we often work not on the group $\mathrm{Ham}(M, \omega)$ itself, but on its universal cover, denoted by $\widetilde{\mathrm{Ham}}(M, \omega)$ (partly for technical, and partly for conceptual reasons). It is convenient to think of an element of

$\widetilde{\text{Ham}}$ as a pair $(\phi, [\alpha])$, where $\phi \in \text{Ham}$ is a Hamiltonian diffeomorphism, $\alpha = \{\alpha_t\}_{t\in[0,1]}$ is a smooth path of Hamiltonian diffeomorphisms with $\alpha_0 = \mathbb{1}$ and $\alpha_1 = \phi$, and $[\alpha]$ stands for the homotopy class of α with fixed endpoints. The universal cover $\widetilde{\text{Ham}}$ is a group with respect to the composition

$$(\phi, [\alpha]) \circ (\psi, [\beta]) = (\phi\psi, [\{\alpha_t\beta_t\}]).$$

The identity element of $\widetilde{\text{Ham}}$, which is still denoted by $\mathbb{1}$, is given by $(\mathbb{1}, [\alpha])$, where α is the constant path.

Exercise 1.4.1. Show that $[\{\alpha_t\beta_t\}] = [\gamma]$, where γ is a concatenation of the paths α_{2t}, $t \in [0, \frac{1}{2}]$ and $\alpha_1\beta_{2t-1}$, $t \in [\frac{1}{2}, 1]$.

Exercise 1.4.2. Let $\{f_t\}_{t\in[0,1]}$ be the Hamiltonian flow generated by a 1-periodic Hamiltonian $F_{t+1} = F_t$. Denote $\tilde{f} = [\{f_t\}_{t\in[0,1]}] \in \widetilde{\text{Ham}}(M, \omega)$. Prove that the element $(\tilde{f})^k$ can be represented by the flow of the Hamiltonian kF_{kt}. *Hint*: Use Proposition 1.3.5 and Exercise 1.4.1.

The fundamental group $\pi_1(\text{Ham})$ is a subgroup of $\widetilde{\text{Ham}}$. Here every homotopy class of loops $\{\gamma_t\}$ with $\gamma_0 = \gamma_1 = \mathbb{1}$ is identified with the element $(\mathbb{1}, [\{\gamma_t\}])$ of $\widetilde{\text{Ham}}$.

Exercise 1.4.3. Show that the center of $\widetilde{\text{Ham}}$ coincides with $\pi_1(\text{Ham})$. Thus, we have a central extension

$$1 \to \pi_1(\text{Ham}) \to \widetilde{\text{Ham}} \to \text{Ham} \to 1.$$

Every Hamiltonian flow $\{f_t\}$, $t \in [0, T]$, with $f_0 = \mathbb{1}$ can be uniquely lifted to a path $\{\tilde{f}_t\}$ on $\widetilde{\text{Ham}}$, where

$$\tilde{f}_t = (f_t, [\{f_{st}\}_{s\in[0,1]}]).$$

Hofer's metric d on Ham lifts to a bi-invariant pseudo-metric $\tilde{d}$ on $\widetilde{\text{Ham}}$ as follows: Given an element $\tilde{\phi} = (\phi, [\alpha]) \in \widetilde{\text{Ham}}$, define

$$\tilde{d}(\mathbb{1}, \tilde{\phi}) = \inf \text{length}\, \beta,$$

where the infimum is taken over all paths $\beta = \{\beta_t\}_{t\in[0,1]}$ with $\beta_0 = \mathbb{1}$, $\beta_1 = \phi$, and $[\beta] = [\alpha]$. By definition, $\tilde{d}(\tilde{a}, \tilde{b}) = \tilde{d}(\mathbb{1}, \tilde{a}^{-1}\tilde{b})$. For many interesting symplectic manifolds one can show that $\tilde{d}$ is nondegenerate; however, this is still unknown in the general case: a priori, $\tilde{d}(\mathbb{1}, \tilde{\phi})$ might vanish for $\tilde{\phi} \in \pi_1(\text{Ham})$. We refer to [105, 106] for a discussion.

1.5. More examples: Kähler manifolds

An important example of symplectic manifolds is given by the Kähler manifolds, that is manifolds equipped with compatible symplectic and complex structures.

Let us first recall some linear algebra. Let E be a complex vector space equipped with a Hermitian inner product $\langle \cdot, \cdot \rangle$. Write

$$\langle u, v\rangle = g(u, v) - \mathrm{i}\omega(u, v).$$

Then $g(\cdot, \cdot)$ is a real inner product, ω an antisymmetric bilinear form, and the two are related by the formula

$$g(u, v) = \omega(u, \mathrm{i}v).$$

For example, in $\mathbb{R}^{2n}(p, q) \simeq \mathbb{C}^n(p + \mathrm{i}q)$ with the standard Hermitian inner product

$$\langle z, w\rangle = \sum_{k=1}^{n} z_k \overline{w}_k$$

we get the standard symplectic form $\omega_0 = \sum_k \mathrm{d}p_k \wedge \mathrm{d}q_k$.

Proposition 1.5.1. *The 2-form ω is nondegenerate.*

Proof. Let $\xi \neq 0$. We must find η such that $\omega(\xi, \eta) \neq 0$. Take $\eta = \mathrm{i}\xi$. Then

$$\omega(\xi, \eta) = \omega(\xi, \mathrm{i}\xi) = g(\xi, \xi) > 0. \qquad \square$$

Definition 1.5.2. Let M^{2n} be a smooth manifold. An *almost complex structure* on M is a smooth field of automorphisms

$$J_x \colon T_xM \to T_xM$$

such that

$$J_x^2 = -\mathbb{1}.$$

An almost complex structure J is called *integrable* (or simply a *complex structure*) if M has the structure of a complex manifold (i.e., a smooth manifold with holomorphic transition maps $\mathbb{C}^n \to \mathbb{C}^n$) so that J is the multiplication by $\mathrm{i} = \sqrt{-1}$.

The *Newlander–Nirenberg theorem* [115] states that the almost complex structure J is integrable if and only if the *Nijenhuis tensor* $N_J \colon TM \otimes TM \to TM$,

$$N_J(\xi, \eta) := [\xi, \eta] + J[J\xi, \eta] + J[\xi, J\eta] - [J\xi, J\eta],$$

vanishes. Here ξ, η are local extensions of the vectors ξ and η to vector fields on M, and $[\cdot, \cdot]$ stands for the commutator. Observe that $N_J(\xi, \xi) = N_J(\xi, J\xi) = 0$. If $\dim M = 2$ and $\xi \neq 0$, the tangent space to M is generated by ξ and $J\xi$ and therefore N_J vanishes identically. We conclude that *every almost complex structure on a surface is integrable.* Furthermore, if $M = S^2$, the uniformization theorem of complex analysis states that all complex structures on M are diffeomorphic. Therefore the sphere S^2 carries a canonical (up to a diffeomorphism) almost complex structure.

Exercise 1.5.3. Prove Smale's theorem [150]: the group $\mathrm{Diff}_+(S^2)$ of orientation preserving diffeomorphisms of S^2 retracts to $\mathrm{SO}(3)$. *Hint*: Denote by $\mathcal{J}$ the space of all almost complex structures on S^2 compatible with the orientation. By the above discussion, $\mathrm{Diff}_+(S^2)$ acts transitively on $\mathcal{J}$. Since $\mathcal{J}$ is a contractible space (why?), $\mathrm{Diff}_+(S^2)$ retracts to the stabilizer of the standard complex structure $j \in \mathcal{J}$, that is to the group of Möbius transformations $\mathrm{PSL}(2, \mathbb{C})$. The latter retracts to $\mathrm{SO}(3)$.

Let M be a manifold equipped with a symplectic form ω and an almost complex structure J. We say that ω and J are *compatible* if $\omega(\cdot, J\cdot)$ defines a Riemannian metric.

Definition 1.5.4. A Kähler manifold is a manifold M^{2n} equipped with a symplectic form ω and a compatible complex structure J.

A basic example of a Kähler manifold is the complex projective space $\mathbb{C}P^n$ equipped with the Fubini–Study metric, described in Section 1.7 below. Furthermore, any complex submanifold of a Kähler manifold is necessarily Kähler with respect to the induced symplectic and complex structures (prove this!). In particular, every complex projective manifold is Kähler.

1.6. J-holomorphic curves

The theory of J-holomorphic curves, introduced by Gromov in his famous paper [76], is an important tool in studying symplectic manifolds. In this section we briefly introduce J-holomorphic curves, which will be used many times throughout this book. For a comprehensive treatment of the subject we refer the reader to the book [108], as well as the lecture notes [160].

Recall from Section 1.5 that an almost complex structure J on M is said to be *compatible* with ω if the bilinear form $\omega(\cdot, J\cdot)$ is a Riemannian metric. We refer the reader to [107] for preliminaries on compatible almost complex structures.

Exercise 1.6.1. Let J be a compatible almost complex structure on M, and let g be the corresponding Riemannian metric. Show that for every smooth function H on M its gradient is given by $\nabla H = -J \,\mathrm{sgrad}\, H$.

Given an almost complex structure J on M, the tangent bundle TM becomes a complex vector bundle and hence one can define its first Chern class $c_1 = c_1(TM, J)$ (see [30] or [110] for details about the first Chern class). The space of all compatible almost complex structures is known to be contractible, and in particular connected. Therefore, c_1 does not depend on the specific choice of a compatible J. In what follows we refer to $c_1(TM, J)$ as to the first Chern class of M and focus on its restriction to spherical classes only. In other words c_1 will stand for the functional $\pi_2(M) \to \mathbb{Z}$.

In general, almost complex manifolds do not carry holomorphic functions. It was a great insight of Gromov [76] that on symplectic manifolds

(M, ω) equipped with a compatible almost complex structure one can develop a theory of *curves* which to a large extent resembles classical algebraic geometry of curves in complex manifolds. Let (Σ, j) be a real surface equipped with an almost complex (and hence, complex, see Section 1.5) structure j. We call (Σ, j) a *Riemann surface.* A map $u\colon (\Sigma, j) \to (M, J)$ is called *J-holomorphic* if $u_* \circ j = J \circ u_*$. Geometrically, if u is J-holomorphic and $u_{*z} \neq 0$ for some $z \in \Sigma$, the tangent space $T_{u(z)}u(\Sigma) \subset T_{u(z)}M$ is invariant under J.

Definition 1.6.2. A J-holomorphic map from a Riemann surface to M is called a J-holomorphic (or pseudoholomorphic) curve.[2]

Exercise 1.6.3. Let $\mathbb{C}$ be the complex line equipped with the complex coordinate $x + \mathrm{i}y$. Prove that a map $u\colon (\mathbb{C}, \mathrm{i}) \to (M, J)$ is J-holomorphic if and only if

$$\frac{\partial u}{\partial x} + J\frac{\partial u}{\partial y} = 0. \tag{1.4}$$

Note that in the case when $(M, J) = (\mathbb{C}, \mathrm{i})$ this becomes the usual Cauchy–Riemann equation.

More generally, any J-holomorphic curve $u\colon (\Sigma, j) \to (M, J)$ satisfies (1.4) in local complex coordinates $z = x + \mathrm{i}y$ on Σ.

Exercise 1.6.4. Prove that for any nonconstant J-holomorphic curve $u\colon \Sigma \to M$

$$\int_\Sigma u^*\omega > 0.$$

Hint: Use the local form (1.4) to show that the integrand $u^*\omega$ is positive near a regular point $z \in \Sigma$ and is nonnegative everywhere.

1.7. Marsden–Weinstein reduction

Let us present a fundamental construction called the Marsden–Weinstein reduction [100], which in particular yields the above-mentioned Fubini–Study symplectic form on $\mathbb{C}P^n$. Let H be a time-independent Hamiltonian function on a symplectic manifold (M, ω) generating a Hamiltonian flow $\{h_t\}$.

Exercise 1.7.1 (Energy conservation law). Prove that $H \circ h_t = H$.

Therefore, M is foliated by the level sets of H, which are invariant under the flow $\{h_t\}$.

Exercise 1.7.2. Fix an energy level $\Sigma = \{H = c\}$, and take any point $x \in \Sigma$ such that $\operatorname{sgrad} H(x) \neq 0$. Show that the vector $\operatorname{sgrad} H(x)$ generates the ω-orthogonal subspace

$$T_x\Sigma^\omega = \{\xi \in T_xM : \omega(\xi, \eta) = 0 \ \forall \eta \in T_x\Sigma\}.$$

---

[2]The terminology *curve* comes from the fact that a Riemann surface is a one-dimensional complex manifold.

Assume now in addition that the flow h_t is T-periodic, i.e., $h_T = \mathbb{1}$ for some $T > 0$. In other words, h_t defines a Hamiltonian S^1-action on M. Suppose that Σ is a regular level set of H on which the action is free. Then the quotient manifold $N^{2n-2} = \Sigma/S^1$ carries a natural symplectic form σ, which is uniquely determined by the fact that $\pi^*\sigma = \omega|_\Sigma$, where $\pi\colon \Sigma \to N$ is the natural projection. Explicitly, let $x \in N$, and let $\xi, \eta \in T_xN$. Choose $\tilde{x} \in \pi^{-1}(x)$ and $\tilde{\xi}, \tilde{\eta} \in T_{\tilde{x}}\Sigma$ such that

$$\pi_*\tilde{\xi} = \xi, \quad \pi_*\tilde{\eta} = \eta.$$

Put

$$\sigma(\xi, \eta) = \omega(\tilde{\xi}, \tilde{\eta}).$$

We claim that the form σ is well defined. Observe that the vectors $\tilde{\xi}, \tilde{\eta}$ are defined up to addition of vectors from $\ker \pi_*$. The line $\ker \pi_*$ is tangent to the fiber $\pi^{-1}(x)$, and hence it is generated by $\operatorname{sgrad} H$. By Exercise 1.7.2, $\operatorname{sgrad} H$ lies in the kernel of $\omega|_\Sigma$. Therefore $\omega(\tilde{\xi}, \tilde{\eta})$ does not depend on the specific choice of the lifts $\tilde{\xi}$ and $\tilde{\eta}$. Furthermore, any two points $\tilde{x}, \tilde{y} \in \pi^{-1}(x)$ are related by $\tilde{y} = h_t\tilde{x}$ for some t. Since h_t is a symplectomorphism, one readily checks that $\sigma(\xi, \eta)$ does not depend on the lift $\tilde{x} \in \pi^{-1}(x)$. The claim follows.

We leave it as an exercise to show that the 2-form σ is closed and non-degenerate, and hence is a symplectic form on N. We refer to (N, σ) as the *reduced* symplectic manifold.

Example 1.7.3 (Harmonic oscillator). Consider $\mathbb{R}^{2n} \simeq \mathbb{C}^n(p + \mathrm{i}q)$, and take

$$H(p, q) = \pi \sum_{k=1}^{n} (p_k^2 + q_k^2).$$

Then the level set

$$\Sigma := \{H = 1\}$$

is the sphere S^{2n-1} of radius $1/\sqrt{\pi}$ centered at the origin. Hamilton's equations

$$\begin{cases} \dot{p}_k &= -2\pi q_k, \\ \dot{q}_k &= 2\pi p_k, \end{cases}$$

can be rewritten in the complex coordinates as

$$\dot{z}_k = 2\pi \mathrm{i} z_k.$$

Therefore, the Hamiltonian flow h_t defines the standard circle action

$$z \mapsto \mathrm{e}^{2\pi \mathrm{i} t} z$$

on Σ, and the reduced manifold is the complex projective space $\mathbb{C}P^{n-1}$. The resulting symplectic form σ on $\mathbb{C}P^{n-1}$ is called the Fubini–Study form. One readily checks that the complex structure induced from $\mathbb{C}^n$ is compatible with σ, which makes $\mathbb{C}P^{n-1}$ a Kähler manifold.

Exercise 1.7.4. Show that for the projective line $\mathbb{C}P^1 \subset \mathbb{C}P^{n-1}$

$$\int_{\mathbb{C}P^1} \sigma = 1.$$

Exercise 1.7.5. Identify

$$S^2 = \{|z|^2 + x^2 = 1\} \subset \mathbb{C} \times \mathbb{R}.$$

Define $\Phi\colon (\mathbb{C}P^1, \sigma) \to (S^2, 1/(4\pi)\Omega)$ by

$$[z_1 : z_2] \mapsto \left(\frac{2z_1\bar{z}_2}{|z_1|^2 + |z_2|^2}, \frac{|z_2|^2 - |z_1|^2}{|z_1|^2 + |z_2|^2} \right).$$

(Here σ is the Fubini–Study form on $\mathbb{C}P^1$ and Ω the induced Euclidean area form on S^2). Prove that Φ is a symplectomorphism.

CHAPTER 2

C^0-Rigidity of the Poisson Bracket

In this chapter we discuss the C^0-rigidity of the Poisson bracket. We prove that the functional

$$(F, G) \mapsto \|\{F, G\}\|$$

is lower semicontinuous in the C^0-topology. As an application, we present a proof of the Eliashberg–Gromov Rigidity Theorem [52, 77], which asserts that the group of compactly supported symplectomorphisms of a symplectic manifold (M, ω) is C^0-closed in the group of all compactly supported diffeomorphisms of M. Finally, we review similar rigidity phenomena related to iterated Poisson brackets.

2.1. Rigidity of the Poisson bracket

The Poisson bracket of two smooth functions is defined via their first derivatives, and thus a priori one does not expect that it should behave robustly under C^0-perturbations of the functions. However, such robustness does hold. This is the content of our main result in this section. We denote by $C_c^\infty(M)$ the space of compactly supported smooth functions on M and write $\|F\|$ for the uniform norm $\max|F|$.

Theorem 2.1.1 ([64]). *The functional*

$$C_c^\infty(M) \times C_c^\infty(M) \to [0, \infty), \quad (F, G) \mapsto \|\{F, G\}\|$$

is lower semicontinuous in the C^0-topology:

$$\liminf_{\overline{F}, \overline{G} \xrightarrow{C^0} F, G} \|\{\overline{F}, \overline{G}\}\| = \|\{F, G\}\| \tag{2.1}$$

for all $F, G \in C_c^\infty(M)$.

This theorem extends earlier results by Cardin–Viterbo [44] and Zapolsky [163]. Note that genuine continuity of the Poisson bracket cannot be hoped for, since one can arbitrarily increase the value of $\|\{F, G\}\|$ by arbitrarily small C^0-perturbations of F and G.

The rigidity of the Poisson bracket allows us to extend the notion of Poisson commutativity to *continuous* functions: we say that two continuous functions F, G Poisson commute, and write $\{F, G\} \equiv 0$, if there exist sequences $F_k, G_k \in C^\infty(M)$ such that

$$F_k \xrightarrow{C^0} F, \quad G_k \xrightarrow{C^0} G \quad \text{and} \quad \{F_k, G_k\} \xrightarrow{C^0} 0.$$

Theorem 2.1.1 ensures that for $F, G \in C_c^\infty(M)$, this definition coincides with the usual notion of Poisson commutativity.

In this section we present two quite different proofs of Theorem 2.1.1. The first one follows [64] and uses results on minimality of geodesics in Hofer's geometry. The second, shorter proof, due to Buhovsky [32], is based on positivity of Hofer's displacement energy of open displaceable subsets.

2.1.1. The first proof (via Hofer's geodesics). Here we present the first proof of Theorem 2.1.1, borrowed from [64], and then outline a refinement of this theorem, estimating the 'rate of convergence' in (2.1) (Theorem 2.1.7 below). We start with the case when M is a closed symplectic manifold. Necessary adjustments for open symplectic manifolds are discussed at the end of this section.

The crucial ingredient of the proof is the following deep result regarding Hofer's metric $\tilde{d}$ on the group $\widetilde{\mathrm{Ham}}(M,\omega)$. As in Section 1.4, we write $\tilde{f}_t$ for the lift of the Hamiltonian flow f_t to $\widetilde{\mathrm{Ham}}$.

Theorem 2.1.2. *Let (M,ω) be a closed symplectic manifold. Let $\{f_t\}$ be a one-parameter subgroup of* $\mathrm{Ham}(M,\omega)$ *generated by an autonomous Hamiltonian $F \in \mathcal{H}$. Then there exists $\varepsilon > 0$, depending on F, such that*

$$\tilde{d}(\mathbb{1}, \tilde{f}_t) = \mathrm{length}\{f_t\} = t\|F\| \quad \forall t \in (0,\varepsilon).$$

Thus one-parameter subgroups of Hamiltonian diffeomorphisms are globally minimizing geodesics on sufficiently small time intervals. Under certain additional assumptions on F, this result was outlined in [132, Chapter 13] and proved in [118] by Floer-homological methods. The complete proof is due to McDuff [103, Proposition 1.5(i)], who deduced the result from a generalized nonsqueezing theorem by using techniques introduced by Lalonde and McDuff in [96]. McDuff's work involves theory of J-holomorphic curves.

We now turn to the proof of Theorem 2.1.1, which we break into several steps.

Lemma 2.1.3. *Let $\{a_t\}, \{b_t\}$ be the Hamiltonian flows of the normalized Hamiltonians A_t, B_t. Then*

$$\tilde{d}(\tilde{a}_1, \tilde{b}_1) \le \int_0^1 \|A_t - B_t\|\, dt.$$

PROOF. The flow $\{a_t^{-1} b_t\}$ is generated by the Hamiltonian

$$-A \circ a_t + B \circ a_t = (B_t - A_t) \circ a_t.$$

Therefore,

$$\tilde{d}(\tilde{a}_1, \tilde{b}_1) = \tilde{d}(\mathbb{1}, \tilde{a}_1^{-1}\tilde{b}_1) \le \mathrm{length}\{a_t^{-1} b_t\}$$
$$= \int_0^1 \|(B_t - A_t) \circ a_t\|\, dt = \int_0^1 \|B_t - A_t\|\, dt,$$

where in the last equality we used the fact that the uniform norm is invariant under composition with diffeomorphisms. □

Given a group, we denote by $[x, y]$ the commutator of its elements x and y: $[x, y] = xyx^{-1}y^{-1}$.

Lemma 2.1.4. *Let $\{f_t\}$, $\{g_t\}$ be the Hamiltonian flows of the autonomous Hamiltonians $F, G \in \mathcal{H}$. Then*

$$\tilde{d}(\mathbb{1}, [\tilde{f}_t, \tilde{g}_t]) \le t^2 \|\{F, G\}\|.$$

PROOF. By the time rescaling property (Proposition 1.3.5), it is enough to prove the lemma for $t = 1$. Denote $g = g_1$, and consider the path $\{f_s g f_s^{-1} g^{-1}\}$, $s \in [0,1]$, which is generated by $H_s := F - F \circ g^{-1} \circ f_s^{-1}$. We leave it as an exercise to show that this path is homotopic to $[f_s, g_s]$, $s \in [0,1]$, with fixed endpoints, and hence represents the element $[\tilde{f}_1, \tilde{g}_1] \in \widetilde{\mathrm{Ham}}$. Now,

$$\begin{aligned} \|H_s\| &= \|H_s \circ f_s\| = \|F \circ f_s - F \circ g^{-1}\| = \|F - F \circ g^{-1}\| \\ &= \left\| \int_0^1 \frac{\mathrm{d}}{\mathrm{d}u} (F \circ g_u^{-1})\, \mathrm{d}u \right\| = \left\| \int_0^1 \{G, F\} \circ g_u^{-1}\, \mathrm{d}u \right\| \le \|\{G, F\}\|. \end{aligned}$$

Here we used the fact that $F \circ f_s \equiv F$, by Exercise 1.7.1. Therefore,

$$\tilde{d}(\mathbb{1}, [\tilde{f}_1, \tilde{g}_1]) \le \mathrm{length}\{f_s g f_s^{-1} g^{-1}\} = \int_0^1 \|H_s\|\, \mathrm{d}s \le \|\{F, G\}\|. \qquad \square$$

The next lemma is the key step in the proof, which uses Theorem 2.1.2 above on geodesics in Hofer's geometry (see formula (2.3) below).

Lemma 2.1.5. *Let F, G, f_t, g_t as before. Then there exist constants $K = K(F, G)$ and $\varepsilon = \varepsilon(F, G)$ such that*

$$\tilde{d}(\mathbb{1}, [\tilde{f}_t, \tilde{g}_t]) \ge t^2 \|\{F, G\}\| - Kt^3$$

for all $0 < t < \varepsilon$.

PROOF. Set $h_t = [f_t, g_t]$. Fix t, and consider the flow $\{f_{st} g_t f_{st}^{-1} g_t^{-1}\}$, $s \in [0, 1]$, which is generated by the Hamiltonian $H^{(t)}(x, s) = t\, U(t)$, where

$$U(t) = F - F \circ g_t^{-1} \circ f_{st}^{-1}.$$

As above, this path is homotopic with fixed endpoints to the path $[f_{st}, g_{st}]$, $s \in [0, 1]$, and hence represents the element $[\tilde{f}_t, \tilde{g}_t] \in \widetilde{\mathrm{Ham}}$. Let us compute the first terms in the Taylor expansion of $H^{(t)}$ as a function of t:

$$\begin{aligned} &H^{(0)} = 0, \\ &\left.\frac{\mathrm{d}}{\mathrm{d}t}\right|_{t=0} H^{(t)} = [U(t) + tU'(t)]\Big|_{t=0} = 0, \\ &\left.\frac{\mathrm{d}^2}{\mathrm{d}t^2}\right|_{t=0} H^{(t)} = [2U'(t) + tU''(t)]\Big|_{t=0} = 2U'(0) \\ &\qquad = -2\big[\{G, F\} \circ g_t^{-1} \circ f_{st}^{-1} + s\{F, F \circ g_t^{-1}\} \circ f_{st}^{-1}\big]\Big|_{t=0} = 2\{F, G\}. \end{aligned}$$

We get that

$$H^{(t)} = \{F, G\} t^2 + O(t^3).$$

Therefore, there exists K such that for small t,

$$\|H^{(t)} - t^2\{F,G\}\| \le Kt^3. \tag{2.2}$$

Denote by φ_τ the Hamiltonian flow generated by $\{F,G\}$. By Theorem 2.1.2, for small t,

$$\tilde{d}(\mathbb{1}, \tilde{\varphi}_{t^2}) = t^2\|\{F,G\}\|. \tag{2.3}$$

On the other hand, φ_{t^2} and $[f_t, g_t]$ are the time-one maps of the Hamiltonian flows of the Hamiltonians $t^2\{F,G\}$ and $H^{(t)}$, respectively, and hence from Lemma 2.1.3 and (2.2) we obtain

$$\tilde{d}([\tilde{f}_t, \tilde{g}_t], \tilde{\varphi}_{t^2}) \le \|H^{(t)} - t^2\{F,G\}\| \le Kt^3. \tag{2.4}$$

Formulas (2.3) and (2.4) together with the triangle inequality yield

$$\tilde{d}(\mathbb{1}, [\tilde{f}_t, \tilde{g}_t]) \ge t^2\{F,G\} - Kt^3.$$

□

Finally, we will need the following elementary result:

Exercise 2.1.6. Consider a group equipped with a bi-invariant pseudo-metric ρ. Prove that for any elements a, b, c, d

$$\rho(\mathbb{1}, [a,b]) - \rho(\mathbb{1}, [c,d]) \le 2[\rho(a,c) + \rho(b,d)].$$

We are now ready to prove our main result.

Proof of Theorem 2.1.1. Fix F, G, and take any functions A, B such that

$$\|F - A\|, \|G - B\| \le \delta, \tag{2.5}$$

where $\delta > 0$ is small enough. Denote by f_t, g_t, a_t, b_t the Hamiltonian flows of F, G, A, B, respectively. By Lemma 2.1.4,

$$\tilde{d}(\mathbb{1}, [\tilde{a}_t, \tilde{b}_t]) \le t^2\|\{A,B\}\|.$$

By Lemma 2.1.5,

$$\tilde{d}(\mathbb{1}, [\tilde{f}_t, \tilde{g}_t]) \ge t^2\|\{F,G\}\| - Kt^3$$

for $K = K(F,G)$ and $0 < t < \varepsilon(F,G)$. Note that, crucially, K and ε do not depend on A, B. Now, by (2.5) and Lemma 2.1.3,

$$\tilde{d}(\tilde{f}_t, \tilde{a}_t) \le \delta t, \quad \tilde{d}(\tilde{g}_t, \tilde{b}_t) \le \delta t,$$

and hence from Exercise 2.1.6 we obtain

$$|\tilde{d}(\mathbb{1}, [\tilde{f}_t, \tilde{g}_t]) - \tilde{d}(\mathbb{1}, [\tilde{a}_t, \tilde{b}_t])| \le 2[\tilde{d}(\tilde{f}_t, \tilde{a}_t) + \tilde{d}(\tilde{g}_t, \tilde{b}_t)] \le 4\delta t.$$

Combining all the above yields

$$t^2\|\{A,B\}\| \ge t^2\|\{F,G\}\| - Kt^3 - 4\delta t.$$

Dividing by t^2 we obtain

$$\|\{A,B\}\| \ge \|\{F,G\}\| - \left[Kt + \frac{4\delta}{t}\right].$$

Finally, let us balance this inequality: choosing $t = \sqrt{\delta}$, we get

$$\|\{A, B\}\| \geq \|\{F, G\}\| - C(F, G)\sqrt{\delta} \tag{2.6}$$

for some constant $C(F, G)$. This completes the proof of the theorem. □

In fact, one can estimate the rate of convergence. For this, we introduce the following definition: Denote

$$\Phi(F, G) = \|\{F, G\}\| \tag{2.7}$$

and

$$\Phi_\delta(F, G) = \inf_{\substack{\|F-\overline{F}\| \leq \delta \\ \|G-\overline{G}\| \leq \delta}} \left\|\{\overline{F}, \overline{G}\}\right\|. \tag{2.8}$$

The following result was first proved by L. Buhovsky [32].

Theorem 2.1.7 (Buhovsky's 2/3 law). *There exists $\delta_0(F, G)$ such that for all $\delta \in (0, \delta_0)$,*

$$\Phi(F, G) - \Phi_\delta(F, G) \leq C \cdot \Psi(F, G)^{1/3} \cdot \delta^{2/3}, \tag{2.9}$$

where $C \leq 100$ is a constant independent of F and G, and

$$\Psi(F, G) = \left\|\Big\{\{\{F, G\}, F\}, F\Big\} + \Big\{\{\{F, G\}, G\}, G\Big\}\right\|.$$

Moreover, for generic F and G, the power $\delta^{2/3}$ is sharp.

In [61] this result was reproved by using Hofer's geometry. Here we give a sketch of the proof.

Outline of the proof for closed manifolds. The main idea is that one can refine Lemma 2.1.5, and obtain

$$\tilde{d}(\mathbb{1}, [\tilde{f}_t, \tilde{g}_t]) \geq t^2\|\{F, G\}\| - Kt^4. \tag{2.10}$$

This is done as follows. Denote by φ_τ the Hamiltonian flow generated by $\{F, G\}$. As before, by Theorem 2.1.2, for small t, $\tilde{d}(\mathbb{1}, \tilde{\varphi}_{t^2}) = t^2\|\{F, G\}\|$. The innovation is the introduction of the 'twisting factor' θ_t, the Hamiltonian flow generated by the Hamiltonian $(F + G)/2$. One can show, by a long computation, that

$$\tilde{d}(\tilde{\varphi}_{t^2}, \tilde{\theta}_t[\tilde{f}_t, \tilde{g}_t]\tilde{\theta}_t^{-1}) \leq Kt^4,$$

for some constant K. By the bi-invariance of Hofer's norm,

$$\tilde{d}(\mathbb{1}, [\tilde{f}_t, \tilde{g}_t]) = \tilde{d}(\mathbb{1}, \tilde{\theta}_t[\tilde{f}_t, \tilde{g}_t]\tilde{\theta}_t^{-1}).$$

These estimates are summarized in Figure 2.1. Thus the triangle inequality gives this time

$$\tilde{d}(\mathbb{1}, [\tilde{f}_t, \tilde{g}_t]) \geq t^2\|\{F, G\}\| - K \cdot \Psi(F, G)t^4,$$

which is (2.10). Then, proceeding as in the proof of Theorem 2.1.1, we get

$$\|\{A, B\}\| \geq \|\{F, G\}\| - \left[Kt^2 + \frac{4\delta}{t}\right].$$

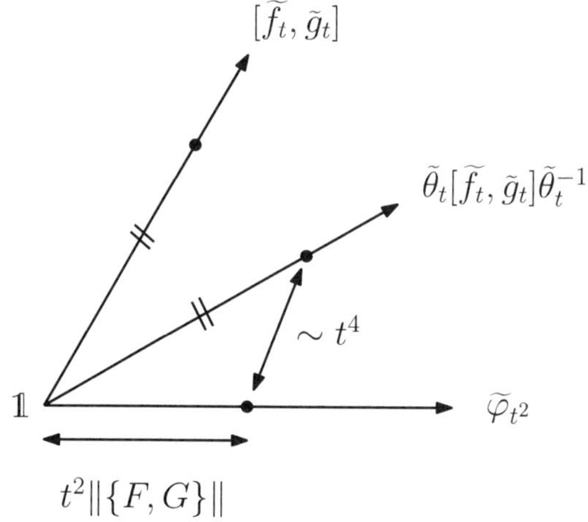

FIGURE 2.1. Buhovsky's 2/3 law

Taking this time $t \sim \delta^{1/3}$ yields the desired result. □

Now let us turn to the case when M is an open symplectic manifold. We start with the discussion of the length-minimizing property of one-parameter subgroups with respect to Hofer's metric. For the standard symplectic vector space $M = (\mathbb{R}^{2n}, \mathrm{d}p \wedge \mathrm{d}q)$ a statement similar to Theorem 2.1.2 was proved by Hofer in [86, Theorem 3] (and, historically, this was the very first statement on Hofer's geodesics). For general open symplectic manifolds, one has the following slightly weaker version of Theorem 2.1.2 (see [106, Proposition 1.7]).

Theorem 2.1.8. *Let (M, ω) be a closed symplectic manifold. Let $\{f_t\}$ be a one-parameter subgroup of* $\mathrm{Ham}(M, \omega)$ *generated by an autonomous Hamiltonian $F \in \mathcal{H}$. Fix any open subset $U \subset M$ with compact closure which contains the support of F. Then there exists $\varepsilon > 0$, depending on F and U, such that for every $t \in [0, \varepsilon)$, for any Hamiltonian path γ joining the identity with f_t, homotopic to $\{f_s\}$, $s \in [0, t]$, and whose Hamiltonian is supported in U*

$$\mathrm{length}(\gamma) \geq t\|F\|.$$

The key difference between Theorems 2.1.2 and 2.1.8 is that in the latter the length-minimizing time interval $[0, \varepsilon)$ depends on the domain U. In order to understand the impact of this nuance (which was missed in [61, 64]) on the proof of Theorems 2.1.1 and 2.1.7, take any compactly supported functions F and G on M. For every domain $U \subset M$ with compact closure which contains the supports of F and G and for every $\delta > 0$, define

$$\Phi_{\delta,U}(F, G) := \inf \|\{\overline{F}, \overline{G}\}\|,$$

where the infimum is taken over all $\overline{F}, \overline{G} \in C_c^\infty(M)$ which are supported in U and which satisfy

$$\|F - \overline{F}\| \le \delta, \quad \|G - \overline{G}\| \le \delta.$$

Note that $\Phi_{\delta,U}(F,G) \ge \Phi_\delta(F,G)$, where $\Phi_\delta(F,G)$ is given by formula (2.8). Arguing exactly as in the closed case above, we get that there exists $\delta_0 = \delta_0(F,G,U)$ (mind the dependence on U) such that for every $\delta \in (0, \delta_0)$

$$\Phi(F,G) - \Phi_{\delta,U}(F,G) \le C(F,G,U) \cdot \delta^{1/2}, \tag{2.11}$$

as in (2.6), and

$$\Phi(F,G) - \Phi_{\delta,U}(F,G) \le C \cdot \Psi(F,G)^{1/3} \cdot \delta^{2/3}, \tag{2.12}$$

as in (2.9). Our goal is to eliminate the dependence on U, that is to replace $\Phi_{\delta,U}(F,G)$ in these inequalities by $\Phi_\delta(F,G)$. This is done with the help of the following lemma.

Lemma 2.1.9. *For all $\delta > 0$*

$$\Phi_\delta(F,G) \ge \Phi_{3\delta,U}(F,G). \tag{2.13}$$

Clearly, the lemma combined with (2.11) and (2.12) yields inequalities (2.6) and (2.9) up to the multiplicative constants $3^{1/2}$ and $3^{2/3}$ in their right-hand sides respectively, and hence completes the proof of Theorems 2.1.1 and 2.1.7 for open symplectic manifolds.

Proof of Lemma 2.1.9. Take a smooth function $v\colon \mathbb{R} \to \mathbb{R}$ such that $v(s) = 0$ for $s \in [-\delta, \delta]$ and $|v(s) - s| \le 2\delta$, $|v'(s)| \le 1$ for all s. Observe that for every function $\overline{F}$ with $\|F - \overline{F}\| \le \delta$, the restriction of $|\overline{F}|$ to $M \setminus U$ is $\le \delta$. Hence, the function $v(\overline{F})$ is supported in U. Furthermore,

$$\|F - v(\overline{F})\| \le \|F - \overline{F}\| + \|\overline{F} - v(\overline{F})\| \le 3\delta.$$

The same inequalities hold for G and $\overline{G}$. Note now that

$$\|\{v(\overline{F}), v(\overline{G})\}\| = \|v'(\overline{F}) \cdot v'(\overline{G}) \cdot \{\overline{F}, \overline{G}\}\| \le \|\{\overline{F}, \overline{G}\}\|.$$

It follows that

$$\Phi_{3\delta,U}(F,G) \le \inf_{\substack{\|F-\overline{F}\|\le\delta \\ \|G-\overline{G}\|\le\delta}} \|\{v(\overline{F}), v(\overline{G})\}\| \le \inf_{\substack{\|F-\overline{F}\|\le\delta \\ \|G-\overline{G}\|\le\delta}} \|\{\overline{F}, \overline{G}\}\| = \Phi_\delta(F,G),$$

as required. □

2.1.2. The second proof (via Hofer's displacement energy). We present in this section another proof, due to Buhovsky [32], of Theorem 2.1.1. The "hard" symplectic ingredient of the proof is positivity of Hofer's displacement energy $e_H(U)$ of every displaceable open subset U of a symplectic manifold (M, ω) (see discussion at the end of Section 1.3.4 above). We start with the following lemma.

Let $U \subset M$ be an open displaceable subset with displacement energy $e_H(U) = r > 0$. Assume that for some $s > 0$ the time-s map g_s of the

Hamiltonian flow generated by a normalized autonomous Hamiltonian G on M displaces U: $g_s U \cap U = \varnothing$.

Lemma 2.1.10. *For every pair of normalized functions $A, B \in C_c^\infty(M)$ with*

$$\|A - G\| < r/s \tag{2.14}$$

and

$$\inf_{g_s(U)} B - \sup_U B > q > 0, \tag{2.15}$$

one has

$$\|\{A, B\}\| > q/s. \tag{2.16}$$

PROOF. Write d for Hofer's metric on $\mathrm{Ham}(M, \omega)$. Denote by a_t the Hamiltonian flow of A. Arguing as in Lemma 2.1.3, we get from (2.14) that

$$d(\mathbb{1}, g_s^{-1} a_s) = d(a_s, g_s) < s \cdot (r/s) = r.$$

Since $e_H(U) = r$, we have that $g_s^{-1} a_s(U) \cap U \neq \varnothing$, which we rewrite as $a_s(U) \cap g_s(U) \neq \varnothing$. This means[1] that there exists a point $x \in U$ such that $a_s x \in g_s(U)$. By (2.15),

$$q < B(a_s x) - B(x) = \int_0^s \{B, A\}(a_t x)\, \mathrm{d}t \leq \|\{A, B\}\| \cdot s,$$

which proves (2.16). □

PROOF OF THEOREM 2.1.1. Let $F, G \in C_c^\infty(M)$ be any pair of functions with

$$\|\{F, G\}\| = K > 0.$$

We have to check that for every $\delta > 0$ there exists $\epsilon > 0$, such that for all $\overline{F}, \overline{G} \in C_c^\infty(M)$ with

$$\|\overline{F} - F\| < \epsilon, \quad \|\overline{G} - G\| < \epsilon$$

one has

$$\|\{\overline{F}, \overline{G}\}\| > K - 4\delta. \tag{2.17}$$

We shall assume without loss of generality that $\delta < K/4$, $\|\{F, G\}\| = \max\{F, G\}$, and that (in the case when the manifold M is closed) the functions $F, G, \overline{F}, \overline{G}$ have zero mean.

Take a point $x \in M$ with $\{F, G\}(x) = K$. Denote by g_t the Hamiltonian flow of G. Choose $s > 0$ and a neighborhood U of x with the following properties:

(i) $\{F, G\}(g_t y) > K - \delta$ for all $y \in U$ and $t \in [0, s]$;
(ii) $|F(y) - F(z)| < s\delta$ for all $y, z \in U$.

[1] We learned this argument from Jinxin Xue.

These inequalities imply that for all $y, z \in U$

$$\begin{aligned} F(g_s y) - F(z) &= \big(F(g_s y) - F(y)\big) + \big(F(y) - F(z)\big) \\ &= \int_0^s \{F, G\}(g_t y)\, \mathrm{d}t + \big(F(y) - F(z)\big) > s(K - \delta) - s\delta, \end{aligned}$$

and hence

$$F(g_s y) - F(z) > s(K - 2\delta). \tag{2.18}$$

Since $K - 2\delta > 2\delta$, it follows from condition (ii) that $g_s U \cap U = \varnothing$.

Put $r = e_H(U)$. Choose any normalized functions $\overline{F}$ and $\overline{G}$ with

$$\|\overline{F} - F\| < \epsilon, \quad \|\overline{G} - G\| < \epsilon,$$

where

$$0 < \epsilon < \min(r/s, s\delta).$$

Note that by (2.18)

$$\inf_{g_s(U)} \overline{F} - \sup_U \overline{F} > s(K - 4\delta) > 0.$$

Apply now Lemma 2.1.10 with $A = \overline{G}$, $B = \overline{F}$ and $q = s(K - 4\delta)$. By the lemma,

$$\|\{\overline{F}, \overline{G}\}\| \geq q/s = K - 4\delta,$$

and hence we obtain inequality (2.17). This completes the proof. □

2.2. Rigidity of symplectomorphisms

In this section we deduce the Eliashberg–Gromov rigidity theorem from the C^0-rigidity of the Poisson bracket (see [43] for a related argument). Note that in dimension 2 symplectomorphisms are nothing more than area preserving diffeomorphisms, which are known to be C^0-closed in the group of all diffeomorphisms. We will therefore focus on higher dimensions:

Theorem 2.2.1. *Let (M^{2n}, ω) be a symplectic manifold, $n \geq 2$. Then* $\mathrm{Symp}(M, \omega)$ *is C^0-closed in the group of all compactly supported diffeomorphisms of M.*

The proof is divided into several steps. The following proposition is the crux of our argument:

Proposition 2.2.2. *Let $f \in \mathrm{Diff}(M)$ be a diffeomorphism such that $f^*\omega \neq \pm\omega$. Then there exist $F, G \in C_c^\infty(M)$ such that $\|\{F, G\}\| = \frac{1}{2}$, but $\|\{f^*F, f^*G\}\| \geq 1$.*

Here and below $\mathrm{Diff}(M)$ stands for the group of all compactly supported diffeomorphisms of M. The proof of Proposition 2.2.2 will be given later, but first let us use it to deduce the Eliashberg–Gromov theorem:

PROOF OF THEOREM 2.2.1. Let $f_k \in \mathrm{Symp}(M,\omega)$ converge in the C^0-topology to $f \in \mathrm{Diff}(M)$. First we prove that $f^*\omega = c\omega$ for some $c \in \{-1,1\}$. Indeed, if this does not hold, then by Proposition 2.2.2 there exist $F, G \in C_c^\infty(M)$ such that

$$\|\{F,G\}\| = \tfrac{1}{2}, \quad \|\{f^*F, f^*G\}\| \geq 1.$$

The sequences f_k^*F and f_k^*G converge in the C^0-topology to f^*F and f^*G, respectively, and by Exercise 1.3.14,

$$\|\{f_k^*F, f_k^*G\}\| = \|\{F,G\}\| = \tfrac{1}{2}.$$

We get that

$$\liminf_{k\to\infty} \|\{f_k^*F, f_k^*G\}\| = \tfrac{1}{2} < \|\{f^*F, f^*G\}\|,$$

which contradicts the lower semicontinuity of the Poisson bracket. Therefore, $f^*\omega = c\omega$ for $c \in \{-1,1\}$.

Next, let us show that $c = 1$. For this, consider the symplectic manifold $N := M \times \mathbb{T}^2$, with symplectic form $\omega \oplus dp \wedge dq$, where p, q are coordinates on $\mathbb{T}^2$. Consider the sequence

$$\mathrm{Symp}(N) \ni f_k \times \mathbb{1} \xrightarrow{C^0} f \times \mathbb{1} \in \mathrm{Diff}(N).$$

Then, by the above proof,

$$(f \times \mathbb{1})^*(\omega \oplus dp \wedge dq) = c_1(\omega \oplus dp \wedge dq), \quad c_1 = \pm 1.$$

On the other hand,

$$(f \times \mathbb{1})^*(\omega \oplus dp \wedge dq) = f^*\omega \oplus dp \wedge dq = c\omega \oplus dp \wedge dq.$$

Comparing these expressions we find that $c = c_1 = 1$, and the proof is complete. □

Remark 2.2.3. When M is closed, the fact that $c = 1$ can be deduced from the following simpler argument by F. Charette: Since f_k converge uniformly to f, for large k, f_k is homotopic to f. Therefore, $[f^*\omega] = [f_k^*\omega] = [\omega] \in H^2(M;\mathbb{R})$. On the other hand, by the first part of the proof above, $[f^*\omega] = c[\omega]$. Since on a closed manifold $[\omega] \neq 0$, we deduce that $c = 1$.

Let us now turn to the proof of Proposition 2.2.2. This will require some linear algebra.

Let (E,ω) be a symplectic vector space. We denote by $\mathrm{Sp}(E,\omega)$ the group of linear symplectomorphisms of E. When $(E,\omega) = (\mathbb{R}^{2n}, \omega_0)$ we denote this group by $\mathrm{Sp}(2n)$, and call it the *linear symplectic group*. For a detailed study of properties of this group we refer to the reader to Section 2.2 of [107]. Recall the isomorphism

$$I_\omega \colon E \to E^*, \quad \xi \mapsto i_\xi\omega$$

induced by ω. Define a symplectic form ω^* on the dual space E^* by

$$\omega^*(\alpha,\beta) = \omega(I_\omega^{-1}\alpha, I_\omega^{-1}\beta) = \alpha(I_\omega^{-1}\beta).$$

Note that on a symplectic manifold (M,ω),

$$\omega^*(\mathrm{d}F,\mathrm{d}G)=\omega(\operatorname{sgrad}F,\operatorname{sgrad}G)=\{G,F\}.$$

Exercise 2.2.4. Let ω_1, ω_2 be two symplectic forms on M and let $c\in\mathbb{R}\setminus\{0\}$. Prove that

$$\omega_2=c\omega_1 \iff \omega_2^*=\frac{1}{c}\omega_1^*.$$

Lemma 2.2.5. *Let E be a $2n$-dimensional vector space, and let ω_1,ω_2 be symplectic forms on E such that for any $\xi,\eta\in E$,*

$$\omega_1(\xi,\eta)=0 \implies \omega_2(\xi,\eta)=0.$$

Then $\omega_2=c\cdot\omega_1$ for some $c\neq 0$.

PROOF. Since in dimension 2 all nonzero 2-forms are proportional, we assume that $n\geq 2$. Choose a symplectic basis for ω_1, that is, a basis $e_1,f_1,\dots e_n,f_n$ of E such that

$$\omega_1(e_i,e_j)=\omega_1(f_i,f_j)=0,\quad \omega_1(e_i,f_j)=\delta_{ij}.$$

Here $\delta_{ij}=1$ when $i=j$ and $\delta_{ij}=0$ otherwise. Then by the assumption on ω_2

$$\omega_2(e_i,e_j)=\omega_2(f_i,f_j)=0,\quad \omega_2(e_i,f_j)=0 \qquad \text{for } i\neq j.$$

Denote $c_i=\omega_2(e_i,f_i)$. We have

$$\omega_1(e_1+f_2,f_1+e_2)=\omega_1(e_1,f_1)-\omega_1(e_2,f_2)=0,$$

whence

$$0=\omega_2(e_1+f_2,f_1+e_2)=c_1-c_2.$$

Therefore, $c_1=c_2$. Similarly, $c_i=c_j$ for all i,j and thus $\omega_2=c_1\cdot\omega_1$. □

Exercise 2.2.6. Let (E^{2n},ω) be a symplectic vector space with $n\geq 2$. Prove that the map

$$\wedge\,\omega\colon \Lambda^1E\to\Lambda^3E$$

is injective.

Corollary 2.2.7. *Let (M^{2n},ω) be a symplectic manifold, $n\geq 2$, and let ω_1,ω_2 be symplectic forms on M such that for any pair of tangent vectors ξ,η,*

$$\omega_1(\xi,\eta)=0 \implies \omega_2(\xi,\eta)=0.$$

Then $\omega_2=c\cdot\omega_1$ for some $c\neq 0$.

PROOF. By Lemma 2.2.5, there exists a nowhere vanishing function $f\in C^\infty(M)$ such that $\omega_2=f\cdot\omega_1$. Moreover,

$$0=\mathrm{d}\omega_2=\mathrm{d}f\wedge\omega_1,$$

whence, by Exercise 2.2.6, $\mathrm{d}f=0$ and thus f is constant. □

Exercise 2.2.8. Let (E,ω) be a symplectic vector space. Let $\alpha,\beta\in E^*\setminus\{0\}$ such that $\omega^*(\alpha,\beta)=0$. Prove that for any $c\neq 0$ there exists $A\in\operatorname{Sp}(E,\omega)$ such that $A^*\alpha=c\alpha$, $A^*\beta=c\beta$.

Lemma 2.2.9. *Let $B^{2n}(r) \subset \mathbb{R}^{2n}$ be the Euclidean ball of radius $r > 0$, and let $A \in \mathrm{Sp}(2n)$. Then there exists $\varphi \in \mathrm{Ham}(\mathbb{R}^{2n}, \omega_0)$, supported in $B^{2n}(r)$, such that*

$$\varphi(0) = 0, \quad D_0\varphi = A.$$

PROOF. We recall that the group $\mathrm{Sp}(2n)$ is connected and choose a path $A_t \in \mathrm{Sp}(2n)$ such that $A_0 = \mathbb{1}$ and $A_1 = A$. By the same argument as in the proof of Proposition 1.3.7, the flow of symplectomorphisms $\{A_t\}$ is generated by some Hamiltonian $H(x,t)$. Let ρ be a cutoff function such that

$$\rho\colon \mathbb{R}^{2n} \to [0,1], \quad \rho \equiv 0 \text{ on } \mathbb{R}^{2n} \setminus B(r), \quad \text{and} \quad \rho \equiv 1 \text{ in a neighborhood of } 0.$$

Then the Hamiltonian $\widetilde{H} := \rho H$ has support in $B(r)$ and $\widetilde{H} = H$ in a neighborhood of 0. Therefore, denoting by φ_t the flow of $\widetilde{H}$, we have that $\varphi_t(x) = A_t x$ provided x is sufficiently close to the origin. In particular, $\varphi_1(0) = 0$ and $D_0\varphi_1 = A$. Therefore, φ_1 is the required Hamiltonian diffeomorphism. □

Finally, we are ready to prove Proposition 2.2.2.

PROOF OF PROPOSITION 2.2.2. Let $f \in \mathrm{Diff}(M)$ such that $f^*\omega \neq \pm\omega$. Since f is compactly supported and ω^n is a volume form, $f^*\omega \neq c\omega$ for any $c \in \mathbb{R} \setminus \{0\}$ by the degree theorem. We claim that we can find $x \in M$ and $\alpha, \beta \in T_y^*M$, where $y = f(x)$, such that

$$\omega_y^*(\alpha, \beta) = 0 \quad \text{and} \quad \omega_x^*(f^*\alpha, f^*\beta) = 1. \tag{2.19}$$

To see this, note that the forms ω_y and $(f^{-1})^*\omega_x$ are proportional if and only if the forms ω_y^* and $\omega_x^*(f^*\cdot, f^*\cdot)$ are proportional. Hence, our assumption and Corollary 2.2.7 together imply that there exists points $x, y = f(x)$ in M and co-vectors $\alpha_0, \beta_0 \in T_y^*M$ such that $\omega_y^*(\alpha_0, \beta_0) = 0$ and $\omega_x^*(f^*\alpha_0, f^*\beta_0) > 0$. Now, rescaling α_0, β_0 we get $\alpha, \beta \in T_y^*M$ which satisfy (2.19). Next, choose compactly supported smooth functions $\overline{F}, \overline{G}$ such that $\mathrm{d}_y\overline{F} = \alpha$ and $\mathrm{d}_y\overline{G} = \beta$, and denote $K = \|\{\overline{F}, \overline{G}\}\|$. Then there exists $\varphi \in \mathrm{Ham}(M, \omega)$ such that

$$\varphi(y) = y, \quad \varphi^*\alpha = \sqrt{2K}\alpha, \quad \varphi^*\beta = \sqrt{2K}\beta.$$

To construct φ, first find a linear symplectomorphism as in Exercise 2.2.8 for α, β, with $c = \sqrt{2K}$, and then construct φ in local coordinates using Lemma 2.2.9. Finally, set

$$F = \frac{1}{\sqrt{2K}}\varphi^*\overline{F}, \quad G = \frac{1}{\sqrt{2K}}\varphi^*\overline{G}$$

and note that

$$\mathrm{d}_yF = \frac{1}{\sqrt{2K}}\,\mathrm{d}\varphi^*\overline{F} = \frac{1}{\sqrt{2K}}\varphi^*\alpha = \alpha, \quad \mathrm{d}_yG = \frac{1}{\sqrt{2K}}\,\mathrm{d}\varphi^*\overline{G} = \frac{1}{\sqrt{2K}}\varphi^*\beta = \beta,$$

and hence

$$-\{f^*F, f^*G\}(x) = \omega_x^*(f^*\,\mathrm{d}F, f^*\,\mathrm{d}G) = \omega_x^*(f^*\alpha, f^*\beta) = 1.$$

On the other hand,

$$\|\{F,G\}\| = \frac{1}{2K}\|\{\overline{F},\overline{G}\}\| = \frac{1}{2},$$

as desired. This completes the proof of Proposition 2.2.2 and, therefore, of the Eliashberg–Gromov theorem. □

2.3. Higher Poisson brackets

In this section we return to the Poisson bracket, and consider rigidity phenomena related to iterated Poisson bracket. Our first result shows that, in general, there is no such rigidity.

Theorem 2.3.1 ([64]). *For any compactly supported functions $F_1, F_2, F_3 \in C^\infty(\mathbb{R}^2)$ and any $\varepsilon > 0$ there exist functions F_1', F_2', F_3' such that $\|F_i - F_i'\| \leq \varepsilon$ and*

$$\{\{F_1', F_2'\}, F_3'\} \equiv 0.$$

PROOF. Define a *thick grid* with mesh c as a union of pairwise disjoint open squares with side length $2c$ whose centers form a rectangular grid of side length $3c$. It is easy to construct, for any c, thick grids X_1, X_2, X_3 with mesh c that cover $\mathbb{R}^2$. Additionally, for small enough c one can find compactly supported $F_1', F_2', F_3' \in C^\infty(\mathbb{R}^2)$ such that $\|F_i - F_i'\| \leq \varepsilon$ and F_i' is constant on X_i. Then, for any $x \in \mathbb{R}^2$, one of the functions F_i' is constant near x, and hence

$$\{\{F_1', F_2'\}, F_3'\}(x) = 0.$$ □

However, if one restricts to iterated Poisson brackets of only *two* functions, rigidity reappears. To state our results, we introduce the following terminology:

Let $\Phi\colon C_c^\infty(M) \times C_c^\infty(M) \to [0,\infty)$. Set

$$\bar{\Phi}(F,G) = \liminf_{\overline{F},\overline{G}\xrightarrow{C^0}F,G} \Phi(\overline{F},\overline{G}).$$

We say that Φ is *weakly robust* if for any $F, G \in C_c^\infty(M)$,

$$\Phi(F,G) > 0 \implies \bar{\Phi}(F,G) > 0.$$

This definition is motivated by [44]. With this notation, Φ is lower semicontinuous if $\bar{\Phi} = \Phi$.

Theorem 2.3.2 ([61]). (1) *The functional $(F,G) \mapsto \big\|\{\{F,G\},G\}\big\|$ is weakly robust, but not lower semicontinuous in the C^0-topology.*
(2) *The functional $(F,G) \mapsto \big\|\{\{F,G\},G\}\big\| + \big\|\{\{F,G\},F\}\big\|$ is lower semicontinuous in the C^0-topology.*

Let us mention that (2) is proved by using the methods of Hofer's geometry. Here we will only prove the first part of (1). For this, we need the following result:

Exercise 2.3.3 (The Hadamard–Kolmogorov–Landau inequalities).

(1) Prove the classical Hadamard–Kolmogorov–Landau (HKL) inequality: suppose that $u\colon \mathbb{R} \to \mathbb{R}$ is a twice-differentiable function such that $\|u\|$, $\|u'\|$, $\|u''\| < \infty$. Then

$$\|u'\| \le C \cdot \|u\|^{1/2} \cdot \|u''\|^{1/2},$$

where C is a numerical constant.

(2) Use the classical HKL inequality to prove its symplectic counterpart obtained in [61]: for any compactly supported functions $F, G \in C_c^\infty(M)$,

$$\|\{F, G\}\| \le C \cdot \|F\|^{1/2} \cdot \big\|\{\{F, G\}, G\}\big\|^{1/2}.$$

Proof of weak robustness. Consider $F, G \in C_c^\infty(M)$ such that $\big\|\{\{F, G\}, G\}\big\| > 0$. Note that this implies that $\{F, G\} \not\equiv 0$. Then the symplectic HKL inequality and the C^0-rigidity of the Poisson bracket yield

$$\begin{aligned} \liminf_{\overline{F},\overline{G}\xrightarrow{C^0}F,G} \big\|\{\{\overline{F}, \overline{G}\}, \overline{G}\}\big\| &\ge \liminf_{\overline{F},\overline{G}\xrightarrow{C^0}F,G} C^{-2} \cdot \frac{\|\{\overline{F}, \overline{G}\}\|^2}{\|\overline{F}\|} \\ &\ge C^{-2}\frac{\|\{F, G\}\|^2}{\|F\|} > 0. \end{aligned}$$ □

Let us remark that using a generalization of the HKL inequality for higher derivatives, the same proof generalizes for higher functionals of the form

$$(F, G) \mapsto \Big\|\Big\{\dots\{\{F, G\}, G\}, \dots G\Big\}\Big\|.$$

However, in general the question of weak robustness or semicontinuity of higher Poisson brackets is still widely open. For instance, it is still unknown whether the functional

$$(F, G) \mapsto \Big\|\Big\{\Big\{\{\{F, G\}, G\}, F\Big\}, G\Big\}\Big\|$$

is weakly robust or semicontinuous. For further discussion see [61].

CHAPTER 3

Quasi-morphisms

In this chapter we introduce quasi-morphisms of groups, a notion from geometric group theory, which plays an important role in this book.

3.1. Homomorphisms up to a bounded error

Roughly speaking, group quasi-morphisms are function to $\mathbb{R}$ which are homomorphisms up to a bounded error.

Definition 3.1.1. Let G be a group. A function $\mu\colon G \to \mathbb{R}$ is called a *quasi-morphism* if there exists a constant $C \geq 0$ such that

$$|\mu(gh) - \mu(g) - \mu(h)| \leq C \quad \forall g, h \in G.$$

A quasi-morphism is called *homogeneous* if in addition

$$\mu(g^n) = n \cdot \mu(g) \quad \forall g \in G, n \in \mathbb{Z}.$$

The smallest constant C in the definition above is called the *defect* of μ. Note that a quasi-morphism with defect zero is simply a homomorphism. The following proposition shows that given any quasi-morphism μ, there exists a homogeneous quasi-morphism $\overline{\mu}$ such that the difference $\mu - \overline{\mu}$ is a bounded function on G. We leave it as an exercise to show that $\overline{\mu}$ is uniquely defined.

Proposition 3.1.2. *Let $\mu\colon G \to \mathbb{R}$ be a quasi-morphism with defect C. Set*

$$\overline{\mu}(g) = \lim_{n\to\infty} \frac{\mu(g^n)}{n}.$$

Then

(i) *The limit exists for all $g \in G$.*
(ii) *$\overline{\mu}$ is a homogeneous quasi-morphism.*
(iii) *$|\overline{\mu} - \mu| \leq C$.*

We shall refer to $\overline{\mu}$ as to *the homogenization* of μ.

PROOF. We first prove that the limit exists. Let $g \in G$. Note that

$$n\mu(g) - (n-1)C \leq \mu(g^n) \leq n\mu(g) + (n-1)C. \tag{3.1}$$

This inequality is symmetric with respect to replacing μ by $-\mu$. Therefore, assume without loss of generality that $\mu(g) \geq 0$ and set $a_n = \mu(g^n)+(n+1)C$. Then by (3.1),

$$a_n \geq n\mu(g) + 2C \geq 0.$$

Further,

$$a_{m+n} = \mu(g^{m+n}) + (m+n+1)C \leq \mu(g^n) + \mu(g^m) + C + (m+n+1)C$$
$$= a_m + a_n.$$

Therefore, a_n is a nonnegative subadditive sequence, and by Fekete's lemma, the limit $\lim_{n\to\infty} a_n/n$ exists. Hence the limit

$$\lim_{n\to\infty} \frac{\mu(g^n)}{n} = \lim_{n\to\infty} \left(\frac{a_n}{n} - \frac{n+1}{n} C \right)$$

exists as well. The proof of homogeneity of $\overline{\mu}$ is left as an exercise. Note that (iii) implies that $\overline{\mu}$ is a quasi-morphism, and so it remains to prove (iii). Note that

$$|\mu(g^n) - n\mu(g)| \leq (n-1)C \implies \left| \frac{\mu(g^n)}{n} - \mu(g) \right| \leq C.$$

Passing to the limit as $n \to \infty$ we get that

$$|\overline{\mu}(g) - \mu(g)| \leq C.$$

□

Remark 3.1.3. It follows from Proposition 3.1.2(iii) that the defect of $\overline{\mu}$ is at most $4C$. See [42, Corollary 2.59] for a better bound.

An important property of homogeneous quasi-morphisms is that they restrict to genuine homomorphisms on abelian subgroups.

Proposition 3.1.4. *Let $\mu\colon G \to \mathbb{R}$ be a homogeneous quasi-morphism. For any $g, h \in G$ such that $gh = hg$, we have $\mu(gh) = \mu(g) + \mu(h)$.*

PROOF.

$$\mu(gh) = \frac{\mu((gh)^n)}{n} = \frac{\mu(g^n h^n)}{n} = \frac{\mu(g^n) + \mu(h^n) + O(1)}{n}$$
$$= \mu(g) + \mu(h) + O\left(\frac{1}{n}\right) \xrightarrow[n\to\infty]{} \mu(g) + \mu(h).$$

□

Exercise 3.1.5. Prove that any homogeneous quasi-morphism μ is a class function, that is

$$\mu(ghg^{-1}) = \mu(h) \quad \forall g, h \in G.$$

Quasi-morphisms originated in a number of mathematical disciplines such as the theory of maps close to representations (see Shtern's survey [148], where quasi-morphisms are called *pseudocharacters*) and bounded cohomology theory (see Monod's ICM-2006 talk [113]). Quasi-morphisms serve as a powerful tool in the study of the commutator length on groups. We refer to Calegari's book [42] for recent developments in this area. In the past decade, a lot of attention has been focused on quasi-morphisms on transformation groups and their universal covers (see the ICM-2006 talk by Ghys [72]). To mention a few examples, quasi-morphisms have been constructed on (the universal covers of) diffeomorphism groups [31, 71], contactomorphism groups [19, 28, 73, 74], and groups of Hamiltonian diffeomorphisms ([56, 58, 137, 138, 147], see also Section 4.8).

These quasi-morphisms have numerous interesting applications to geometry and dynamics. Ironically, perhaps the most famous quasi-morphism, the Poincaré rotation number on the universal cover of the homeomorphism group of the circle, had been discovered almost a century before the term quasi-morphism was coined. We will discuss this quasi-morphism in Section 3.3. In the next chapter we shall introduce a quasi-morphism on $\widetilde{\mathrm{Ham}}(M, \omega)$, which plays a crucial role in function theory on symplectic manifolds.

3.2. Quasi-morphisms and irreversible dynamics

Quasi-morphisms serve as substitutes of homomorphisms for many geometric and dynamical purposes. Let us give a specific example with a dynamical flavor. Assume that the group G acts on a set X, so that each element $g \in G$ gives rise to a dynamical system $\{g^n\}$ on X with discrete time $n \in \mathbb{Z}$. The dynamics determined by g is called *reversible* if $g = hg^{-1}h^{-1}$ for some $h \in G$. Intuitively, reversibility means that one cannot distinguish the past of the system from its future. Quasi-morphisms can be used for detecting irreversibility: It follows from Exercise 3.1.5 above that the dynamics is irreversible provided $\mu(g) \neq 0$ for some homogeneous quasi-morphism μ. Of course, a homomorphism $G \to \mathbb{R}$ would serve this purpose just as well. The problem is that many interesting groups do not admit nontrivial homomorphisms to $\mathbb{R}$.

For instance, the modular group $\mathrm{PSL}(2, \mathbb{Z})$ is generated by torsion elements and hence does not admit nontrivial homomorphisms to $\mathbb{R}$. Nevertheless, this group admits plenty of homogeneous quasi-morphisms. Let us discuss this issue in the more general context of discrete subgroups of isometries of the hyperbolic half-plane

$$\mathbb{H} = \{z = x + \mathrm{i}y \in \mathbb{C} : y > 0\},$$

equipped with the hyperbolic metric $\mathrm{d}s^2 = y^{-2}(\mathrm{d}x^2 + \mathrm{d}y^2)$. The group of hyperbolic isometries is $\mathrm{PSL}(2, \mathbb{R})$, acting by Möbius transformations $z \mapsto (az + b)/(cz + d)$.

The following construction is due to Barge and Ghys [15]. Let $G \subset \mathrm{PSL}(2, \mathbb{R})$ be a discrete subgroup. Denote by Ω the hyperbolic area form on $\mathbb{H}$. Take any 1-form α on $\mathbb{H}$ which is G-invariant (that is $g^*\alpha = \alpha$ for all $g \in G$) and which satisfies $|\mathrm{d}\alpha/\Omega| \leq C$ for some $C > 0$ (for example, one can take the lift of any compactly supported 1-form on $\mathbb{H}/G$). Fix a base point $z \in \mathbb{H}$. Define $\mu \colon G \to \mathbb{R}$ as follows. Let $[z, az]$ be the geodesic segment in $\mathbb{H}$ between z and az. Set

$$\mu(a) = \int_{[z,az]} \alpha.$$

Proposition 3.2.1. *The function $\mu\colon G \to \mathbb{R}$ is a quasi-morphism.*

PROOF. Let $a, b \in G$. Denote by Δ be the geodesic triangle with vertices z, az, abz. Note that since G acts on $\mathbb{H}$ by isometries and α is G-invariant,

$$\int_{[z,bz]} \alpha = \int_{[az,abz]} \alpha.$$

Therefore,

$$\begin{aligned} |\mu(ab) - \mu(a) - \mu(b)| &= \left| \int_{[z,abz]} \alpha - \int_{[z,az]} \alpha - \int_{[az,abz]} \alpha \right| \\ &= \left| \int_{\partial\Delta} \alpha \right| = \left| \int_{\Delta} \mathrm{d}\alpha \right| \leq C \cdot \mathrm{Area}_{\Omega}(\Delta) \leq C\pi, \end{aligned}$$

since the hyperbolic area of any geodesic triangle does not exceed π. □

Exercise 3.2.2. Prove that the homogenization $\overline{\mu}$ of μ does not depend on the choice of the base point z.

Exercise 3.2.3. Assume that $\Sigma := \mathbb{H}/G$ is a closed hyperbolic surface. Let ℓ be an embedded closed geodesic on Σ, which corresponds to the deck transformation $g \in G$ (here g is uniquely defined up to a conjugation in G). Take any 1-form β on Σ with $\int_\ell \beta \neq 0$. Prove that $\overline{\mu}(g) \neq 0$. If in addition ℓ is homologous to zero, g lies in the commutator subgroup $[G, G]$. In this case $\overline{\mu}$ is a genuine quasi-morphism (that is, not a homomorphism).

Exercise 3.2.4 ([135]). Let $G = \mathrm{PSL}(2, \mathbb{Z})$ and let $h \in G$ be the parabolic element $h(z) = z + 1$. Consider a horoball $B = \{y > R\}$, where $R > 0$ is large enough.

(1) Prove that for $g \in G$, $gB \cap B \neq \varnothing$ if and only if g is a power of h.

(2) Denote by H the cyclic group generated by h and by S the collection of all cosets Hg, $g \in G$. Consider a 1-form $\beta = u(y)\,\mathrm{d}x$ on $\mathbb{H}$, where u vanishes for $y < R$ and $u = 1$ for $y > R + 1$. Since h preserves β, the form $g^*\beta$ does not depend on the representative g of the coset $[g] \in S$. Therefore, the 1-form

$$\alpha = \sum_{[g] \in S} g^*\beta$$

is well defined and satisfies the assumption of the Barge–Ghys construction. Show that $\overline{\mu}(h) = 1$ for the corresponding homogeneous quasi-morphism $\overline{\mu}$.

Let us return to the general case of a discrete $G \subset \mathrm{PSL}(2, \mathbb{R})$. In fact, one can show that if $g \in G$ is any primitive element (that is $g \neq h^k$ for $h \in G$, $|k| > 1$) of infinite order which is not conjugate to its inverse, an appropriate choice of the 1-form α in the construction above yields a homogeneous quasi-morphism $\overline{\mu}$ with $\overline{\mu}(g) \neq 0$ (we refer to [135] for a simple proof and generalizations of this statement). Thus, in the context of discrete groups of hyperbolic isometries, *quasi-morphisms completely detect irreversibility.* This is a manifestation of the fact that the Barge–Ghys construction yields an infinite-dimensional space of quasi-morphisms on the modular group.

3.3. The Poincaré rotation number

Many interesting transformation groups (and/or their universal covers) are perfect, that is, they coincide with their commutator subgroup. As a consequence, they do not admit nontrivial homomorphisms to $\mathbb{R}$. A basic example (see [51, Theorem 2.3]) is given by the universal cover G of the group $\mathrm{Diff}_0(S^1)$ of orientation preserving diffeomorphisms of the circle S^1. It can be identified with

$$G = \{f \in C^\infty(\mathbb{R}) : f \text{ is strictly increasing, } f(x+1) = f(x) + 1 \ \forall x \in \mathbb{R}\}.$$

For $f \in G$, the *Poincaré rotation number* is defined as

$$\mathrm{rot}(f) = \lim_{n\to\infty} \frac{f^n(0)}{n}.$$

Proposition 3.3.1. *The function* $\mathrm{rot}\colon G \to \mathbb{R}$ *is a homogeneous quasi-morphism.*

PROOF. Define $c\colon G \to \mathbb{R}$ by $c(f) = f(0)$. Observe that it suffices to check that c is a quasi-morphism, since rot is the homogenization of c. So, let $f, g \in G$. Denote by $\lceil \cdot \rceil$ the ceiling function. We have

$$f\big(g(0)\big) \le f(\lceil g(0) \rceil) = f(0) + \lceil g(0) \rceil \le f(0) + g(0) + 1.$$

Similarly,

$$f\big(g(0)\big) \ge f(0) + g(0) - 1.$$

Therefore,

$$|c(fg) - c(f) - c(g)| = \big|f\big(g(0)\big) - f(0) - g(0)\big| \le 1,$$

proving that c is a quasi-morphism. □

Example 3.3.2. The group $\mathrm{PSL}(2,\mathbb{R})$ acts on the two-dimensional hyperbolic disc $\mathbb{D}$ by hyperbolic isometries, and the action extends to an action on $\partial\mathbb{D} = S^1$. We get an embedding

$$\mathrm{PSL}(2,\mathbb{R}) \hookrightarrow \mathrm{Diff}_0(S^1),$$

which induces a map of the universal covers

$$\widetilde{\mathrm{PSL}}(2,\mathbb{R}) \xrightarrow{\varphi} \widetilde{\mathrm{Diff}}_0(S^1).$$

The pullback $\varphi^* \mathrm{rot}\colon \widetilde{\mathrm{PSL}}(2,\mathbb{R}) \to \mathbb{R}$ is a quasi-morphism. It lifts to a quasi-morphism $\widetilde{\mathrm{SL}}(2,\mathbb{R}) \to \mathbb{R}$, called *the Maslov quasi-morphism.*

3.4. The Maslov quasi-morphism

The definition of the Maslov quasi-morphism on $\widetilde{\mathrm{SL}}(2, \mathbb{R})$, which appeared at the end of the previous section, can be in fact extended to the universal cover of the linear symplectic group in arbitrary dimension. While this extension can be defined in many ways, we use an approach based on the Robbin–Salamon index of a path of Lagrangian subspaces of $(\mathbb{R}^{2n}, \omega_0)$ (see [139]). We begin by reviewing this invariant.

Definition 3.4.1. A subspace $L \subset (\mathbb{R}^{2n}, \omega_0)$ is called *Lagrangian* if $\dim L = n$ and $\omega_0|_L = 0$.

Remark 3.4.2. Note that due to the nondegeneracy of ω_0, the dimension of a subspace on which ω_0 vanishes is at most n. In general, subspaces on which the symplectic form vanishes are called *isotropic*. Thus Lagrangian subspaces are maximal isotropic subspaces.

Exercise 3.4.3. Let B be an $n \times n$ matrix. Prove that

$$\operatorname{graph} B = \{(p, q) : q = Bp\} \subset (\mathbb{R}^{2n}, \omega_0)$$

is Lagrangian if and only if B is symmetric.

Exercise 3.4.4. Consider $\mathbb{R}^{2n} \times \mathbb{R}^{2n}$ with the split symplectic form $(-\omega_0) \oplus \omega_0$. Prove that the graph of any symplectic linear transformation is a Lagrangian subspace.

We denote by $\Lambda = \Lambda(\mathbb{R}^{2n})$ the Grassmannian manifold of all Lagrangian subspaces of $\mathbb{R}^{2n}$. For $L \in \Lambda$ we identity the tangent space $T_L\Lambda$ with the space of all quadratic forms on L, as follows. Let $L(t)$ be a path in Λ with $L(0) = L$. Write $L(t) = \theta(t)L$ with $\theta(t) \in \mathrm{Sp}(2n)$ and $\theta(0) = \mathbb{1}$. We associate to $\dot{L}(0)$ the quadratic form on L

$$\omega\big(v, \dot{\theta}(0)v\big), \quad v \in L.$$

Exercise 3.4.5. Prove that this quadratic form is well defined, that is, it does not depend on the specific choice of the path of symplectic matrices $\theta(t)$.

Fix $L \in \Lambda$. Define, for $1 \le k \le n$, $\Sigma_L^k = \{E \in \Lambda : \dim(E \cap L) = k\}$ and $\Sigma_L = \bigcup_{k \ge 1} \Sigma_L^k$.

Exercise 3.4.6. Show that Σ_L^k is a smooth submanifold of codimension $k(k+1)/2$ in Λ.

The set Σ_L, consisting of all Lagrangian subspaces which intersect L nontransversally, has a natural structure of a stratified manifold. In view of Exercise 3.4.6, the top stratum Σ_L^1 has codimension 1 in Λ. It carries a natural coorientation defined as follows. Let $E \in \Sigma_L^1$ and let $\xi \in T_E\Lambda$. By our identification, ξ is a quadratic form on E. Then ξ is transversal to $T_E\Sigma_L^1$ if and only if the form $\xi|_{E \cap L}$ is nondegenerate. Since $\dim(E \cap L) = 1$, this form is either positive or negative definite. We say that ξ is positively oriented if $\xi|_{E\cap L}$ is positive definite. Applying again Exercise 3.4.6, we see that the higher strata Σ_L^k, $k \geq 2$ have codimension ≥ 3 in Λ. Therefore Σ_L with the above coorientation defines a singular cycle in Λ, called *the Maslov–Arnold cycle.*

Now we are ready to define the Robbin–Salamon index of a smooth path $\alpha\colon [a,b] \to \Lambda$ of Lagrangian subspaces. If the endpoints of α do not lie on the Maslov–Arnold cycle Σ_L and α is transversal to its strata, the Robbin–Salamon index of α is simply the intersection index of α with Σ_L^1. In order to handle the general case, let us introduce the following terminology. A vector $\xi \in T_E\Lambda$ is called *regular* if $\xi_{E\cap L}$ is nondegenerate. A *crossing* is a parameter value $t \in [a,b]$ such that $\alpha(t) \in \Sigma_L$. We call α *regular* if $\dot\alpha(t)$ is regular for all crossings. For a regular path, the number of crossings is finite. One can show that every path in Λ is homotopic with fixed endpoints to a regular path.

Definition 3.4.7. Let $\alpha\colon [a,b] \to \Lambda$ be a regular path. Its *Robbin–Salamon index* is defined as

$$r_L(\alpha) = \frac{1}{2}\operatorname{sign}\dot\alpha(a)|_{\alpha(a)\cap L} + \sum_{a<t<b} \operatorname{sign}\dot\alpha(t)|_{\alpha(t)\cap L} + \frac{1}{2}\operatorname{sign}\dot\alpha(b)|_{\alpha(b)\cap L},$$

where the sum is taken over all crossings $t \in [a,b]$. Here $\operatorname{sign} Q$ denotes the signature of a quadratic form Q (that is the number of positive squares minus the number of negative squares in the canonical form of Q).

The Robbin–Salamon index has the following properties:

(1) If α is homotopic to β with fixed endpoints, then $r_L(\alpha) = r_L(\beta)$.
(2) $r_L(\alpha \,\#\, \beta) = r_L(\alpha) + r_L(\beta)$, where $\#$ stands for concatenation of paths.
(3) For any $A \in \mathrm{Sp}(2n)$, $r_{AL}(A\alpha) = r_L(\alpha)$.

Proofs of these properties can be found in [139]. Note that the first property allows us to define the Robbin–Salamon index of an arbitrary path as the index of any regular path homotopic to it. Moreover, since every symplectic vector space is linearly symplectomorphic to $(\mathbb{R}^{2n}, \omega_0)$, the third property allows us to define the Robbin–Salamon index of a path of Lagrangian subspaces in any symplectic vector space.

Example 3.4.8. In $\mathbb{R}^2(p,q) \simeq (\mathbb{C}, z = p + \mathrm{i}q)$, take $L = \{q = 0\} \simeq \mathbb{R}$ and $\alpha(t) = \mathrm{e}^{\mathrm{i}t}L$ for $t \in [0,\pi]$. The only crossings are at $t = 0, \pi$. Since

$$\frac{\mathrm{d}}{\mathrm{d}t}\mathrm{e}^{\mathrm{i}t} = \mathrm{i}\mathrm{e}^{\mathrm{i}t},$$

the tangent vectors $\dot\alpha(0)$ and $\dot\alpha(\pi)$ correspond to the quadratic form

$$\omega|_L(z, \mathrm{i}z) = |z|^2 > 0.$$

Therefore, $\operatorname{sign}\dot\alpha(0) = \operatorname{sign}\dot\alpha(\pi) = 1$ and $r_L(\alpha) = \frac{1}{2} + \frac{1}{2} = 1$.

Proposition 3.4.9. *Consider a Lagrangian subspace $L = \operatorname{graph} B \subset \mathbb{R}^{2n}$, where $B = B^T$ is a symmetric matrix. Let $\alpha\colon [a,b] \to \Lambda$ be a regular path of Lagrangian subspaces of the form*

$$\alpha(t) = \operatorname{graph} A(t), \quad A(t) = A(t)^T, \qquad t \in [a,b].$$

Then its Robbin–Salamon index is given by

$$r_L(\alpha) = \tfrac{1}{2}\operatorname{sign}\big(A(1) - B\big) - \tfrac{1}{2}\operatorname{sign}\big(A(0) - B\big).$$

PROOF. Assume for simplicity that $B = 0$ (the general case is left as an exercise). Let us analyse the behavior of $\operatorname{sign} A(t)$ when the parameter t passes through the crossing t_0. Assume without loss of generality that $t_0 = 0$ and $A(0)$ is a diagonal matrix. The argument below is borrowed from [139, pp. 831–832]. Write $A(t)$ in the block form

$$A(t) = \begin{pmatrix} u(t) & v(t) \\ v^T(t) & w(t) \end{pmatrix},$$

where $u(0) = v(0) = 0$ and $w(0)$ is a nondegenerate diagonal matrix. Observe that the restriction of the quadratic form $\dot\alpha(0)$ to $\alpha(0) \cap L$ is given by the quadratic form associated to the symmetric matrix $\dot u(0)$. In a small neighborhood of 0 write $A(t) = P(t)D(t)P^T(t)$, where

$$P(t) = \begin{pmatrix} \mathbb{1} & v(t)w(t)^{-1} \\ 0 & \mathbb{1} \end{pmatrix}, \quad D(t) = \begin{pmatrix} u(t) - v(t)w(t)^{-1}v^T(t) & 0 \\ 0 & w(t) \end{pmatrix}.$$

Differentiating at $t = 0$, we get that

$$D(t) = \begin{pmatrix} t\dot u(0) & 0 \\ 0 & w(0) + t\dot w(0) \end{pmatrix} + O(t^2).$$

Thus for $t > 0$ small enough

$$\operatorname{sign} A(t) = \operatorname{sign} D(t) = \operatorname{sign} w(0) + \operatorname{sign} \dot u(0) = \operatorname{sign} A(0) + \operatorname{sign} \dot\alpha(0)|_{\alpha(0)\cap L}.$$

Similarly, for $t < 0$ with small enough absolute value,

$$\operatorname{sign} A(t) = \operatorname{sign} A(0) - \operatorname{sign} \dot\alpha(0)|_{\alpha(0)\cap L}.$$

Therefore, when the parameter t passes through a crossing t_0, the signature of $A(t)$ changes by $2\operatorname{sign} \dot\alpha(t_0)|_{\alpha(t_0)\cap L}$ for each interior crossing and by $\operatorname{sign} \dot\alpha(t_0)|_{\alpha(t_0)\cap L}$ if the crossing t_0 is one of the endpoints a, b. The proposition follows now from the definition of the Robbin–Salamon index. □

Next, we explore the dependence of the Robbin–Salamon index r_L on the choice of Lagrangian subspace L.

Exercise 3.4.10. (i) Deduce from the properties of the Robbin–Salamon index that for a loop $\alpha\colon S^1 \to \Lambda$, the value of $r_L(\alpha)$ is always an integer which depends only on the homotopy class of α.

(ii) Show that for a loop $\alpha\colon S^1 \to \Lambda$ the value $r_L(\alpha)$ does not depend on the Lagrangian subspace L.

Definition 3.4.11. The Robbin–Salamon index on loops is called *the Maslov index*. The Maslov index gives rise to a generator of the first cohomology group $H^1(\Lambda;\mathbb{Z}) = \mathbb{Z}$ called *the Maslov class* (see [5]).

For general paths (which are not necessarily loops) the dependence of $r_L(\alpha)$ on L is mild in the following sense.

Proposition 3.4.12. *Let $\alpha\colon [0,1] \to \Lambda$ be a path of Lagrangian subspaces. Then $r_{L_1}(\alpha) - r_{L_2}(\alpha) \leq 2n$ for any $L_1, L_2 \in \Lambda(\mathbb{R}^{2n})$.*

PROOF. It follows from Exercise 3.4.10(ii) that for every $L_1, L_2 \in \Lambda$, the difference $r_{L_1}(\alpha) - r_{L_2}(\alpha)$ depends only on the endpoints of α. Assume without loss of generality that

$$L_1 = \operatorname{graph} B_1, \quad L_2 = \operatorname{graph} B_2, \quad \alpha(0) = \operatorname{graph} A_0, \quad \alpha(1) = \operatorname{graph} A_1.$$

Since we have a freedom in the choice of a path joining $\alpha(0)$ and $\alpha(1)$, take

$$\alpha(t) = \operatorname{graph} A_t, \quad A_t = (1-t)A_0 + tA_1.$$

By Proposition 3.4.9,

$$\begin{aligned} &r_{L_1}(\beta) - r_{L_2}(\beta) \\ &\quad = \tfrac{1}{2}[\operatorname{sign}(A_1 - B_1) - \operatorname{sign}(A_0 - B_1) - \operatorname{sign}(A_1 - B_2) + \operatorname{sign}(A_0 - B_2)]. \end{aligned}$$

Since all the signatures lie between 0 and n, the result follows. □

Now we are ready to introduce the Maslov quasi-morphism on $\widetilde{\mathrm{Sp}}(2n)$. Elements of $\widetilde{\mathrm{Sp}}(2n)$ are homotopy classes with fixed endpoints of paths $\varphi\colon [0,1] \to \mathrm{Sp}(2n)$, $\varphi(0) = \mathbb{1}$. As on $\widetilde{\mathrm{Ham}}$ (see Section 1.4), the product of elements $[\varphi]$ and $[\psi]$ can be defined either as the class of the point-wise product $\{\varphi(t)\psi(t)\}$, or as the class of the concatenation of φ and $\varphi(1)\psi$, which we denote by $\varphi\#\varphi(1)\psi$.

Fix a Lagrangian subspace $L \in \Lambda$. For an element $\varphi = [\{\varphi(t)\}] \in \widetilde{\mathrm{Sp}}(2n)$, set

$$\mu_L(\varphi) = r_L(\{\varphi(t)L\}).$$

Proposition 3.4.13. *The function $\mu_L\colon \widetilde{\mathrm{Sp}}(2n) \to \mathbb{R}$ is a quasi-morphism.*

PROOF. Let $\varphi, \psi \in \widetilde{\mathrm{Sp}}(2n)$. We represent the product $\varphi\psi \in \widetilde{\mathrm{Sp}}(2n)$ by concatenation of paths. Set $\alpha(t) = \varphi(t)L$, $\beta(t) = \psi(t)L$. Then we have

$$\mu_L(\varphi\psi) = r_L(\alpha \# \varphi(1)\beta) = r_L(\alpha) + r_L(\varphi(1)\beta) = r_L(\alpha) + r_{\varphi(1)^{-1}L}(\beta).$$

Therefore,

$$\begin{aligned} |\mu_L(\varphi\psi) - \mu_L(\varphi) - \mu_L(\psi)| &= |r_L(\alpha) + r_{\varphi(1)^{-1}L}(\beta) - r_L(\alpha) - r_L(\beta)| \\ &= |r_{\varphi(1)^{-1}L}(\beta) - r_L(\beta)| \leq 2n, \end{aligned}$$

by Proposition 3.4.12. □

Definition 3.4.14. The homogenization of μ_L is called the *Maslov quasi-morphism*, denoted by $\mu\colon \widetilde{\mathrm{Sp}}(2n) \to \mathbb{R}$.

Remark 3.4.15. By Exercise 3.1.4, the restriction of μ to the abelian subgroup $\pi_1(\mathrm{Sp}(2n)) \subset \widetilde{\mathrm{Sp}}(2n)$ is a homomorphism. We denote it by $\mathrm{Maslov}(\gamma)$, $\gamma \in \pi_1(\mathrm{Sp}(2n))$, and call it the *Maslov index* of γ.

Exercise 3.4.16. Show that the Maslov index of the loop of rotations $\{z \to \mathrm{e}^{2\pi \mathrm{i} t} z\}$, $t \in [0,1]$, in $\mathrm{Sp}(2) = \mathrm{SL}(2,\mathbb{R})$ equals 2. Here we identify $\mathbb{R}^2(p,q) = \mathbb{C}(p + \mathrm{i}q)$ and the symplectic form equals $\mathrm{d}p \wedge \mathrm{d}q$.

Exercise 3.4.17. Show that in the case of $\widetilde{\mathrm{Sp}}(2) = \widetilde{\mathrm{SL}}(2,\mathbb{R})$ Definition 3.4.14 agrees with the one given in Example 3.3.2.

Finally, we define yet another index, introduced by Conley and Zehnder in [49], which plays an important role in Floer theory, see Section 11.4 below. Consider the symplectic vector space $(\mathbb{R}^{2n} \oplus \mathbb{R}^{2n}, (-\omega_0) \oplus \omega_0)$. The diagonal $\Delta \subset \mathbb{R}^{2n} \oplus \mathbb{R}^{2n}$ is a Lagrangian subspace.

Definition 3.4.18. Let $\alpha\colon [0,1] \to \mathrm{Sp}(2n)$ be a path based at the identity, $\alpha(0) = \mathbb{1}$. Define its *Conley–Zehnder index* by

$$\overline{\mathrm{CZ}}(\alpha) = \mu_\Delta\big(\mathbb{1} \times \alpha(t)\big).$$

Exercise 3.4.19. Denote by $\mathfrak{sp}(2n)$ the Lie algebra of $\mathrm{Sp}(2n)$. Take any matrix $A \in \mathfrak{sp}(2n)$ with $\det A \neq 0$. Define, for $t \in [0,\varepsilon]$ with $\varepsilon > 0$ small enough, the path $\alpha(t) = \exp(tA) \in \mathrm{Sp}(2n)$.

(1) Show that the path $\alpha(t)$ is generated by the Hamiltonian $H(x) = \frac{1}{2}\omega(x, Ax)$.
(2) Prove that

$$\overline{\mathrm{CZ}}(\alpha) = \tfrac{1}{2} \operatorname{sign} Q,$$

where Q is the quadratic form $Q(x) = \omega(x, Ax)$. *Hint*: Recall Definition 3.4.7.

Remark 3.4.20. In what follows we shall deal with the index

$$\mathrm{CZ}(\alpha) := n - \overline{\mathrm{CZ}}(\alpha). \tag{3.2}$$

A convenient feature of this convention is that the index of the path α of symplectic matrices generated by a small quadratic Hamiltonian H on $\mathbb{R}^{2n}$ equals the Morse index (that is, the number of negative squares in the canonical form) $\mathrm{ind}(H)$ of H. Indeed, by Exercise 3.4.19,

$$\mathrm{CZ}(\alpha) = n - \tfrac{1}{2} \cdot \big(2n - 2 \cdot \mathrm{ind}(H)\big) = \mathrm{ind}(H).$$

By a slight abuse of terminology, we will call CZ the Conley–Zehnder index.

Exercise 3.4.21. Let $\alpha\colon [0,1] \to \mathrm{Sp}(2n)$ be a path based at the identity, $\alpha(0) = \mathbb{1}$. Show that the parity of $\mathrm{CZ}(\alpha)$ depends only on the endpoint $A = \alpha(1)$ provided the latter does not have 1 as an eigenvalue: $\mathrm{CZ}(\alpha)$ is even if and only if $\det(A - \mathbb{1}) > 0$.

3.5. Quasi-morphisms and invariant pseudo-norms

Recall that a *pseudo-norm* on a group G is a function $\nu\colon G \to [0, +\infty)$ with the following properties:

(i) **(Normalization)** $\nu(\mathbb{1}) = 0$,
(ii) **(Symmetry)** $\nu(\phi) = \nu(\phi^{-1})$,
(iii) **(Triangle inequality)** $\nu(\phi\psi) \le \nu(\phi) + \nu(\psi)$,

for all $\phi, \psi \in G$. A pseudo-norm ν is called *a norm* if it is nondegenerate, that is, if $\nu(\phi) > 0$ for all $\phi \neq \mathbb{1}$. The pseudo-norm is *conjugation invariant* if

$$\nu(\phi\psi\phi^{-1}) = \nu(\psi) \quad \forall \phi, \psi \in G.$$

We refer to [36] for a discussion of conjugation invariant (pseudo-) norms on groups of geometric origin. We say that ν is *discrete* if the positive values of ν do not accumulate at 0, and that ν is *bounded* if $\sup_{\phi \in G} \nu(\phi) < \infty$.

The rest of this section follows [59] and [156]. Let G be a group, and let $c\colon G \to \mathbb{R}$ be a function with the following properties:

(i) **(Conjugation invariance)** $c(\phi\psi\phi^{-1}) = c(\psi)\ \forall \phi, \psi \in G$;
(ii) **(Subadditivity)** $c(\phi\psi) \le c(\phi) + c(\psi)$.

Observe that by (ii), $c(\mathbb{1}) \ge 0$ and

$$c(\phi) + c(\phi^{-1}) \ge c(\mathbb{1}) \ge 0 \quad \forall \phi \in G.$$

Put

$$q(\phi) := c(\phi) + c(\phi^{-1}).$$

One readily checks that the function ν_c on G defined by $\nu_c(\phi) = q(\phi)$ for $\phi \neq \mathbb{1}$ and $\nu_c(\mathbb{1}) = 0$ is a conjugation invariant pseudo-norm. Let us make two observations (item (ii) is borrowed from Usher [156]).

Proposition 3.5.1. (i) *If $c(\mathbb{1}) > 0$ then the pseudo-norm ν_c is discrete and nondegenerate.*
(ii) *The pseudo-norm ν_c is bounded if and only if the function $c\colon G \to \mathbb{R}$ is a quasi-morphism.*

PROOF. By subadditivity $q(\phi) \ge q(\mathbb{1}) = 2c(\mathbb{1})$, which proves (i).

Let us pass to (ii). Assume first that ν_c (and hence q) is bounded. Notice that by the subadditivity

$$c(\phi\psi) - c(\phi) - c(\psi) \le 0$$

and

$$c(\phi\psi) - c(\phi) - c(\psi) \ge \big(c(\psi) - c(\phi^{-1})\big) - c(\phi) - c(\psi) = -q(\phi).$$

Thus, c is a quasi-morphism.

In the opposite direction, if c is a quasi-morphism, then

$$q(\phi) = \big(c(\phi) + c(\phi^{-1}) - c(\phi\phi^{-1})\big) + c(\mathbb{1})$$

is bounded, and hence ν_c is bounded. □

For our future purposes let us look more closely at the homogenization of c even when it is not a quasi-morphism.

Proposition 3.5.2. *For every $\phi \in G$ the limit*

$$\sigma(\phi) := \lim_{k\to\infty} c(\phi^k)/k$$

is well defined.

PROOF. Put $a_k = kc(\phi^{-1}) + c(\phi^k)$. Then $a_k \geq c(\phi^{-k}) + c(\phi^k) \geq 0$ and $a_{k+l} \leq a_k + a_l$. The subadditivity implies that the limit $\lim_{k\to\infty} a_k/k$ is well defined, and hence $\sigma(\phi)$ is well defined. □

The following proposition will be useful in Section 4.6.

Proposition 3.5.3.

$$|\sigma(\phi\psi) - \sigma(\phi) - \sigma(\psi)| \leq \min\big(q(\phi), q(\psi)\big) \tag{3.3}$$

for all $\phi, \psi \in G$.

Of course, the proposition is informative provided $\sigma \neq 0$.

PROOF. We start with the formula

$$(\phi\psi)^k = \theta\phi^k, \quad \text{where } \theta = (\phi\psi\phi^{-1})(\phi^2\psi\phi^{-2})\cdots(\phi^k\psi\phi^{-k}). \tag{3.4}$$

Observe that by subadditivity and conjugation invariance,

$$c(\theta) \leq kc(\psi). \tag{3.5}$$

Moreover, since θ^{-1} can be written as a similar product with ψ replaced by ψ^{-1}, we have similarly

$$c(\theta^{-1}) \leq kc(\psi^{-1}). \tag{3.6}$$

(1) Note that by subadditivity we have

$$c(\psi^{-k}) \leq kc(\psi^{-1}) \quad \text{and} \quad c(\psi^{-k}) \geq c(\mathbb{1}) - c(\psi^k) \geq -c(\psi^k),$$

which gives $kc(\psi^{-1}) \geq -c(\psi^k)$. Hence,

$$kc(\psi) = kq(\psi) - kc(\psi^{-1}) \leq kq(\psi) + c(\psi^k).$$

By (3.4) and (3.5),

$$c\big((\phi\psi)^k\big) \leq c(\theta) + c(\phi^k) \leq kc(\psi) + c(\phi^k) \leq kq(\psi) + c(\psi^k) + c(\phi^k).$$

Dividing by k and passing to the limit as $k \to \infty$ we get that

$$\sigma(\phi\psi) - \sigma(\phi) - \sigma(\psi) \leq q(\psi). \tag{3.7}$$

(2) Note that

$$kc(\psi^{-1}) = -kc(\psi) + kq(\psi) \leq -c(\psi^k) + kq(\psi).$$

By (3.4) and (3.6),

$$c(\phi^k) \leq c(\theta^{-1}) + c\big((\phi\psi)^k\big) \leq kc(\psi^{-1}) + c\big((\phi\psi)^k\big).$$

Combining these inequalities, we get that

$$c(\phi^k) \leq -c(\psi^k) + kq(\psi) + c\big((\phi\psi)^k\big).$$

Rearranging the terms, dividing by k, and passing to the limit as $k \to \infty$ we get that

$$\sigma(\phi\psi) - \sigma(\phi) - \sigma(\psi) \geq -q(\psi).$$

Together with (3.7) this yields

$$|\sigma(\phi\psi) - \sigma(\phi) - \sigma(\psi)| \leq q(\psi).$$

Since c is conjugation invariant, σ is also conjugation invariant, and thus $\sigma(\phi\psi) = \sigma(\psi\phi)$. Repeating the above arguments we get that

$$|\sigma(\phi\psi) - \sigma(\phi) - \sigma(\psi)| \leq q(\phi).$$

Together with the previous inequality this yields (3.3). □

CHAPTER 4

Subadditive Spectral Invariants

The main tool for the theory developed in the subsequent chapters comes from the machinery of *subadditive spectral invariants* on the universal cover $\widetilde{\mathrm{Ham}}(M,\omega)$ of a closed symplectic manifold [119, 144, 157]. These invariants play an important role in the study of the algebraic and Hofer-geometric structure of $\widetilde{\mathrm{Ham}}(M,\omega)$. In this chapter we introduce these invariants axiomatically and discuss their basic features, including a link to quasi-morphisms. Their existence will be established later on, in Chapter 12, by means of Floer theory.

4.1. The Calabi homomorphism

Our first task is to define some basic numerical invariants of elements of $\widetilde{\mathrm{Ham}}(M,\omega)$. Even though our ultimate objective is to study Hamiltonian diffeomorphisms of closed symplectic manifolds, we often have to deal with diffeomorphisms supported in their open subsets. In view of that, we start the exposition with the case of an open symplectic manifold (M,ω).

Let $\phi \in \widetilde{\mathrm{Ham}}(M,\omega)$ be an element generated by a compactly supported Hamiltonian $F_t(x) = F(x,t)$. We define

$$\mathrm{Cal}(\phi) = \int_0^1 \mathrm{d}t \int_M F_t\omega^n.$$

The number $\mathrm{Cal}(\phi)$, which was introduced by Calabi [41], can be thought of as the average energy required to generate ϕ.

Theorem 4.1.1. (i) $\mathrm{Cal}(\phi)$ *is well-defined, that is, it does not depend on the specific choice of the Hamiltonian* F *generating* ϕ.
(ii) $\mathrm{Cal}\colon \widetilde{\mathrm{Ham}}(M,\omega) \to \mathbb{R}$ *is a homomorphism.*

We refer to Cal as the *Calabi homomorphism.* Further details on this issue can be found in [13] and in [107, Section 10.3].

Proof. Let $\{f_{t,s}\}$, $s,t \in [0,1]$, be a smooth two-parametric family of Hamiltonian diffeomorphisms, with $f_{0,s} = \mathbb{1}$ for all s, and which represents ϕ for every fixed s. Let $F_{t,s}$ and $H_{t,s}$ be normalized Hamiltonians such that

$$\frac{\partial}{\partial t} f_{t,s}x = \mathrm{sgrad}\, F_{t,s}(f_{t,s}x) \quad \text{and} \quad \frac{\partial}{\partial s} f_{t,s}x = \mathrm{sgrad}\, H_{t,s}(f_{t,s}x) \qquad \forall x \in M.$$

Observe that

$$H_{0,s} = H_{1,s} = 0. \tag{4.1}$$

Exercise 4.1.2 ([13]). Prove that $\partial/\partial s F_{t,s} = \partial/\partial t H_{t,s} - \{F_{t,s}, H_{t,s}\}$.

It follows from the exercise that

$$\begin{aligned} I(s) := \frac{\mathrm{d}}{\mathrm{d}s}\int_0^1 \mathrm{d}t \int_M F_{t,s}\omega^n &= \int_0^1 \mathrm{d}t \int_M \frac{\partial}{\partial s} F_{t,s}\omega^n \\ &= \int_M \left(\int_0^1 \frac{\partial}{\partial t} H_{t,s}\,\mathrm{d}t\right)\omega^n - \int_0^1 \mathrm{d}t \int_M \{F_{t,s}, H_{t,s}\}\omega^n. \end{aligned}$$

By (4.1),

$$\int_0^1 \frac{\partial}{\partial t} H_{t,s}\,\mathrm{d}t = H_{1,s} - H_{0,s} = 0,$$

and by Exercise 1.3.12,

$$\int_M \{F_{t,s}, H_{t,s}\}\omega^n = 0.$$

Thus $I(s) = 0$ for all s, therefore the integral $\int_0^1 \mathrm{d}t \int_M F_{t,s}\omega^n$ does not depend on the homotopy parameter s. This proves that Cal is well defined.

Next, let us prove that it is a homomorphism. Let f_t, g_t be the Hamiltonian flows generated by normalized Hamiltonians F_t, G_t and representing elements $\phi, \psi \in \widetilde{\mathrm{Ham}}(M,\omega)$, respectively. Then the product $f_t g_t$ is generated by the normalized Hamiltonian $F_t + G_t \circ f_t^{-1}$ and hence

$$\begin{aligned} \mathrm{Cal}(\phi\psi) &= \int_0^1 \mathrm{d}t \int_M F_t\omega^n + \int_0^1 \mathrm{d}t \int_M G_t \circ f_t^{-1}\omega^n \\ &= \int_0^1 \mathrm{d}t \int_M F_t\omega^n + \int_0^1 \mathrm{d}t \int_M G_t\omega^n = \mathrm{Cal}(\phi) + \mathrm{Cal}(\psi). \qquad \square \end{aligned}$$

Remark 4.1.3. In the case when (M,ω) is an *exact* symplectic manifold, that is $\omega = \mathrm{d}\lambda$ for some 1-form λ on M, the Calabi homomorphism descends to a homomorphism $\overline{\mathrm{Cal}}\colon \mathrm{Ham}(M,\omega) \to \mathbb{R}$, see [107, Section 10.3].

4.2. The action spectrum

Another source of numerical invariants of Hamiltonian diffeomorphisms comes from their fixed points. In what follows, we work on a closed symplectic manifold (M,ω).

We start with the following auxiliary definition. The *group of periods* Γ_ω is the abelian subgroup of $\mathbb{R}$ formed by integrals of ω over two-dimensional spheres in M. For instance, $\Gamma_\omega = \{0\}$ for the standard symplectic torus, and Γ_ω is generated by a and b for $(S^2 \times S^2, a\omega_0 \oplus b\omega_0)$, where $a, b \neq 0$ and ω_0 is the standard area form on S^2 of total area 1. In the latter example, Γ_ω is discrete when a and b are dependent over the rationals, and is dense otherwise.

Let $\phi \in \widetilde{\mathrm{Ham}}(M,\omega)$ be an element represented by a Hamiltonian path $\{f_t\}$, $t \in [0,1]$, generated by a normalized Hamiltonian $F(x,t)$. Let y be a fixed point of f_1 corresponding to a contractible closed orbit $\gamma = \{f_t y\}$ of the flow. Existence of such fixed points is guaranteed by a version of Arnold's conjecture (see Section 1.2.2 above). Take any disc D capping γ (that is, a disc whose boundary is γ) up to homotopy with fixed boundary. Define the *action* of the pair (y, D) as

$$\mathcal{A}_F(y,D) = \int_0^1 F(f_t y, t)\,\mathrm{d}t - \int_D \omega.$$

Observe that if D' is another disc capping γ, D and D' may be glued along γ to obtain a 2-sphere, and hence $\mathcal{A}_F(y,D) - \mathcal{A}_F(y,D') \in \Gamma_\omega$. Sometimes it would be convenient to emphasize the dependence of the action on the path $\{f_t\}$ rather than on the Hamiltonian. In this case we write $\mathcal{A}(\{f_t\}, y, D)$ instead of $\mathcal{A}_F(y,D)$.

Let $G(x,t)$ be another normalized Hamiltonian generating ϕ. Take any disc Σ capping the (automatically contractible!) closed orbit $\{g_t y\}$ of the point y under the flow g_t of G.

Proposition 4.2.1. $\mathcal{A}_F(y,D) = \mathcal{A}_G(y,\Sigma) \bmod \Gamma_\omega$.

PROOF. It suffices to verify the claim for contractible loops of Hamiltonian diffeomorphisms, that is, for the case when F generates $\phi = \mathbb{1} \in \widetilde{\mathrm{Ham}}(M,\omega)$ and $G = 0$ (we leave this reduction as an exercise). Let f_t, $t \in S^1 = \mathbb{R}/\mathbb{Z}$, $f_0 = f_1 = \mathbb{1}$, be a loop of Hamiltonian diffeomorphisms homotopic to the identity. Write F_t for the normalized Hamiltonian generating $\{f_t\}$. Denote by $\{f_{t,s}\}$, $s \in [0,1]$, a homotopy between $\mathbb{1}$ and $\{f_t\}$, that is, $f_{t,0} = \mathbb{1}$ and $f_{t,1} = f_t$ for all t.

Fix any point $y \in M$. The map

$$u\colon [0,1] \times S^1 \to M, \quad (s,t) \mapsto f_{s,t} y,$$

defines a spanning disc D for the closed orbit $\gamma = \{f_t y\}$. We shall show that the action $\mathcal{A}_F(y,D)$ vanishes, which will yield the proposition.

To this end, introduce normalized Hamiltonians $F_{t,s}$ and $H_{t,s}$ by

$$\frac{\partial}{\partial t} f_{t,s} x = \operatorname{sgrad} F_{t,s}(f_{t,s} x) \quad \text{and} \quad \frac{\partial}{\partial s} f_{t,s} x = \operatorname{sgrad} H_{t,s}(f_{t,s} x) \qquad \forall x \in M.$$

Here F is 1-periodic in t, $F_{t,0} = 0$ and $F_{t,1} = F_t$ for all t, while $H_{0,s} = H_{1,s} = 0$ for all s. Observe that

$$\int_D \omega = \int_{[0,1]\times S^1} u^*\omega = \int_0^1 \int_0^1 \omega\Big(\frac{\partial}{\partial t} f_{t,s} y, \frac{\partial}{\partial s} f_{t,s} y\Big)\,\mathrm{d}s\,\mathrm{d}t$$
$$= \int_0^1 \int_0^1 \{H, F\}\,\mathrm{d}s\,\mathrm{d}t.$$

By Exercise 4.1.2

$$\frac{\partial}{\partial s} F_{t,s} = \frac{\partial}{\partial t} H_{t,s} - \{F_{t,s}, H_{t,s}\}.$$

Substituting and integrating by parts, we obtain that

$$\int_D \omega = \int_0^1 F_t(f_t y).$$

It follows that

$$\mathcal{A}_F(y, D) = \int_0^1 F_t(f_t y) - \int_D \omega = 0 = \mathcal{A}_G(y, \Sigma),$$

where $G = 0$ and Σ is the constant disc spanning the closed orbit $\{g_t y = y\}$. This completes the proof. □

Definition 4.2.2. The *action spectrum* $\mathrm{spec}(\phi) \subset \mathbb{R}$ of $\phi \in \widetilde{\mathrm{Ham}}(M, \omega)$ is defined as the set of all actions $\mathcal{A}_F(y, D)$, where F is a normalized Hamiltonian generating ϕ, y is a fixed point corresponding to a contractible orbit of the Hamiltonian flow, and D is a capping disc.

An important feature of the action spectrum which will be used below is that $\mathrm{spec}(\phi)$ has empty interior in $\mathbb{R}$ (see [118, Lemma 2.2]). Observe also that $\mathrm{spec}(\phi)$ is obviously Γ_ω-invariant.

The structure of the action spectrum is especially simple for *nondegenerate* elements of $\widetilde{\mathrm{Ham}}(M, \omega)$. Consider a path $\{f_t\}$, $t \in [0, 1]$, of Hamiltonian diffeomorphisms, which is generated by a normalized Hamiltonian F. The element ϕ, represented by this path, is said to be nondegenerate if the graph of f_1 is transversal to the diagonal in $M \times M$ (see Section 1.2.2 above). If ϕ is nondegenerate, then f_1 has a finite number of fixed points, and hence $\mathrm{spec}(\phi)$ consists of finitely many Γ_ω-orbits. Nondegeneracy is generic in the following sense. By using a standard transversality argument, one can show that ϕ becomes nondegenerate after an arbitrarily small C^∞-perturbation of the Hamiltonian F.

Let us motivate the definition of the action. We shall start with the following topological preparations. Let ΛM be the space of all contractible loops $z \colon S^1 \to M$. Consider the following cover of ΛM:

$$\widetilde{\Lambda M} = \{(z, u) \colon z \in \Lambda M, u \colon D^2 \to M, u|_{\partial D^2} = z\} / \sim,$$

where

$$(z, u) \sim (z', u') \iff z = z', \int_{D^2} u^*\omega = \int_{D^2} (u')^*\omega.$$

Let $F(x, t)$ be a normalized Hamiltonian function, which we assume to be 1-periodic in time: $F(x, t+1) = F(x, t)$. Let f_t, $t \in [0, 1]$, be the Hamiltonian flow generated by F. Define the *action functional* $\mathcal{A}_F \colon \widetilde{\Lambda M} \to \mathbb{R}$ by

$$\mathcal{A}_F(z, u) = \int_0^1 F(z(t), t)\, dt - \int_{D^2} u^*\omega.$$

The least action principle of classical mechanics asserts that contractible 1-periodic orbits of f_t correspond to critical points of $\mathcal{A}_F$ on $\widetilde{\Lambda M}$. We leave this statement as an exercise (see also Section 11.2). With this language the action spectrum $\mathrm{spec}(\phi)$, $\phi \in \widetilde{\mathrm{Ham}}(M, \omega)$, is simply the set of critical values of $\mathcal{A}_F$, where F is a normalized Hamiltonian generating $\phi \in \widetilde{\mathrm{Ham}}$. This viewpoint is crucial for Floer theory, a version of Morse–Novikov theory for the action functional $\mathcal{A}_F$ on $\widetilde{\Lambda M}$ which will be discussed in Chapter 11.

Let us remark that in order to define an analogue of the Morse index for critical points of the action functional on general symplectic manifolds, one has to pass to more sophisticated covers of ΛM. However, the cover $\widetilde{\Lambda M}$ suffices for the purposes of the present book.

4.3. Subadditive spectral invariants

In a sense, the action spectrum contains *too much* numerical information about a Hamiltonian diffeomorphism. Nevertheless, after an appropriate selection, the points of the action spectrum give rise to interesting algebraic and/or geometric structures on the group $\widetilde{\mathrm{Ham}}$. Let us present an axiomatic approach to such a selection.

Let (M^{2n}, ω) be a closed symplectic manifold. Denote $\mathcal{G} := \widetilde{\mathrm{Ham}}(M, \omega)$. Write $V := \int_M \omega^n$ for the volume of M. Introduce the notation

$$\langle H \rangle := \frac{1}{V} \cdot \int H\, \omega^n \tag{4.2}$$

for the mean value of a function H on M.

Definition 4.3.1. A function $c\colon \mathcal{G} \to \mathbb{R}$ is called *a subadditive spectral invariant* if it satisfies the following axioms:

(i) **(Conjugation invariance)** $c(\phi\psi\phi^{-1}) = c(\psi)\ \forall \phi, \psi \in \mathcal{G}$;

(ii) **(Subadditivity)** $c(\phi\psi) \leq c(\phi) + c(\psi)$;

(iii) **(Stability)**

$$\int_0^1 \min(F_t - G_t)\, dt \leq c(\phi) - c(\psi) \leq \int_0^1 \max(F_t - G_t)\, dt,$$

provided $\phi, \psi \in \mathcal{G}$ are generated by normalized Hamiltonians F and G, respectively;

(iv) **(Spectrality)** $c(\phi) \in \mathrm{spec}(\phi)$ for all nondegenerate elements $\phi \in \mathcal{G}$.

The function c satisfies properties (i), (ii) from Section 3.5. We keep the notations $q(\phi) = c(\phi) + c(\phi^{-1})$ and $\sigma(\phi) = \lim_{k\to\infty} c(\phi^k)/k$ introduced in that section. We also recall the definition of the pseudo-norm ν_c: $\nu_c(\phi) = q(\phi)$ for all $\phi \neq \mathbb{1}$, and $\nu_c(\mathbb{1}) = 0$. We call ν_c the *spectral pseudo-norm* associated to c.

The following picture, which will be established later, may help the reader to digest these notions. In certain situations the subadditive spectral invariant c is a quasi-morphism on the group $\mathcal{G}$ and the function σ is the corresponding homogeneous quasi-morphism. While $\sigma(\phi)$ remembers spectral information about $\phi \in \mathcal{G}$ only "up to a bounded error," it is more handy for algebraic manipulations. Thus c and σ are algebraic structures on $\mathcal{G}$. At the same time, the function q and the closely related spectral pseudo-norm ν_c

define a geometry on $\mathcal{G}$. In fact ν_c is "almost" a norm, see Proposition 4.6.2 below.

As we shall see in Section 4.7, every closed symplectic manifold (M, ω) admits a subadditive spectral invariant. In what follows, c stands for such an invariant.

Remark 4.3.2.[1] If the group of periods Γ_ω is a discrete subgroup of $\mathbb{R}$, then $c(\phi) \in \text{spec}(\phi)$ *for all* elements $\phi \in \mathcal{G}$. Here is a sketch of the proof (we leave the details as an exercise). Choose a sequence of normalized Hamiltonians $F^{(i)}$, $i \in \mathbb{N}$, which generates nondegenerate elements $\phi_i \in \mathcal{G}$, and which C^∞-converges to a Hamiltonian F of ϕ. By the spectrality axiom, $c(\phi_i) = \mathcal{A}_{F_i}(z_i, D_i)$, where z_i is a contractible 1-periodic orbit of F_i, and D_i is a capping disc. Passing to a subsequence, we can assume that z_i converges as $i \to \infty$ to a 1-periodic orbit z of F. Connect z_i with z by an annulus A_i, so that $\int_{A_i} \omega \to 0$. Look at the sequence of discs $\Sigma_i := D_i \# A_i$, capping z (here $\#$ stands for gluing along the boundary). By the stability axiom, $c(\phi_i) \to c(\phi)$. Furthermore,

$$\int_0^1 F_i(z_i(t), t)\, dt \to \int_0^1 F(z(t), t)\, dt.$$

These observations combined with the discreteness of Γ_ω imply that the sequence $\int_{\Sigma_i} \omega$ stabilizes for sufficiently large i. This readily yields $c(\phi) \in \text{spec}(\phi)$.

Exercise 4.3.3. Deduce from the stability axiom that for all $\phi \in \mathcal{G}$

$$q(\phi) \leq 2c(\mathbb{1}) + 2\tilde{d}(\mathbb{1}, \phi), \tag{4.3}$$

where $\tilde{d}$ stands for Hofer's pseudo-metric on $\mathcal{G}$.

Our next objective is to understand the small scale behavior of the invariants c, q and σ introduced above. Here the small scale is understood in the sense of Section 1.3.4: we focus on Hamiltonian isotopies with displaceable supports. A somewhat tricky consideration below culminates with Theorem 4.3.8 providing various estimates on these invariants in terms of the geometry of supports.

Given a displaceable subset $U \subset M$, define its *spectral displacement energy* $e_c(U)$ as $\inf q(h)$, where the infimum is taken over all $h \in \mathcal{G}$ displacing U. By the above exercise,

$$e_c(U) \leq 2c(\mathbb{1}) + 2e_H(U), \tag{4.4}$$

where e_H stands for Hofer's displacement energy.

Proposition 4.3.4. *Let $\phi, \psi \in \mathcal{G}$ be elements generated by normalized Hamiltonians $F(x,t)$ and $G(x,t)$, $t \in [0,1]$, respectively. Then the stability*

[1] Thanks to Viktor L. Ginzburg for this comment.

property holds:

$$\int_0^1 \min(F_t - G_t)\,dt \le \sigma(\phi) - \sigma(\psi) \le \int_0^1 \max(F_t - G_t)\,dt.$$

PROOF. Let us prove the inequality on the right (the one on the left follows by a similar argument). Write f_t and g_t for the flows generated by F and G, respectively. Choose a monotone function $a\colon [0,1] \to [0,1]$ which equals 0 near 0 and 1 near 1. Consider the flows $\bar{f}_t = f_{a(t)}$ and $\bar{g}_t = g_{a(t)}$ generated by the Hamiltonians $\overline{F}(x,t) = a'(t)F\big(x, a(t)\big)$ and $\overline{G}(x,t) = a'(t)G\big(x, a(t)\big)$, respectively (recall Proposition 1.3.5). Observe that $\overline{F}$ and $\overline{G}$ extend to 1-periodic functions of time on $M \times \mathbb{R}$. Therefore, by Exercise 1.4.2, ϕ^k and ψ^k are generated by the Hamiltonians $k\overline{F}(x,kt)$ and $k\overline{G}(x,kt)$ respectively. Put $A(t) = \max(F_t - G_t)$. It follows from periodicity and the stability axiom that

$$\begin{aligned} c(\phi^k) - c(\psi^k) &\le \int_0^1 \max(k\overline{F}_{kt} - k\overline{G}_{kt})\,dt = k\int_0^{1/k} \max(k\overline{F}_{kt} - k\overline{G}_{kt})\,dt \\ &= k\int_0^{1/k} ka'(kt)A\big(a(kt)\big)\,dt = k\cdot\int_0^1 A(t)\,dt. \end{aligned}$$

Dividing by k and passing to the limit as $k \to \infty$, we get the desired result. □

For an open subset $U \subset M$ denote by $\mathcal{G}(U)$ the subgroup of $\mathcal{G}$ consisting of all elements generated by (in general, not normalized!) Hamiltonians supported in U. Note that $\mathcal{G}(U)$ is the image of the natural morphism $\widetilde{\mathrm{Ham}}(U) \to \widetilde{\mathrm{Ham}}(M) = \mathcal{G}$.

Proposition 4.3.5. *Let $U \subset M$ be an open displaceable subset, and choose any $h = [\{h_t\}] \in \mathcal{G}$ such that $h_1U \cap U = \varnothing$. Let ϕ_t, $t \in [0,1]$, be a Hamiltonian path representing $\phi \in \mathcal{G}(U)$, generated by a Hamiltonian $F_t(x)$ with $\bigcup_t \mathrm{supp}(F_t) \subset U$. Then*

$$c(h\phi) = c(h) - \int_0^1 \langle F_t\rangle\,dt. \tag{4.5}$$

Recall (see (4.2) above) that $\langle F_t\rangle$ stands for the mean value of F_t.

For the proof of Proposition 4.3.5 we shall need the following fact.

Exercise 4.3.6. Let $\{g_t\}$, $t \in [0,1]$, be a Hamiltonian flow. For $t \in [0,1]$, denote $\tilde{g}_t = [\{g_s\}_{s\in[0,t]}]$. Deduce from the stability axiom that the function $t \mapsto c(\tilde{g}_t)$ is continuous.

PROOF OF PROPOSITION 4.3.5. The proof is based on the Schwarz–Ostrover trick ([123, 144], see also [145] for a generalization of this trick).

(1) Recall that we denote by $\mathrm{Fix}(g)$ the set of fixed points of a diffeomorphism g. Put $g_t := h_1\phi_t$. We claim that $\mathrm{Fix}(g_t) = \mathrm{Fix}(h_1)$ for all t, and moreover $\mathrm{Fix}(h_1) \subset M \setminus U$. Indeed, if $x \in U$, then since h_1 displaces U and

$\phi_t x \in U$, we have $h_1 x \notin U$ and $g_t x \notin U$, so x is not a fixed point of either h_1 or g_t. If $x \notin U$ we have that $g_t x = h_1 x$, which yields the claim.

(2) Assume first that h is nondegenerate. Fix t and consider the Hamiltonian path γ obtained by concatenation of the paths $\alpha := \{h_s\}$, $s \in [0,1]$, and $\beta := \{h_1\phi_s\}$, $s \in [0,t]$. Given a fixed point $x \in \mathrm{Fix}(h_1) = \mathrm{Fix}(g_t)$, since $\phi_s x = x$ for all s, the corresponding closed orbit of γ, up to a change of time, coincides with $\{h_s x\}$. Assume that this orbit is contractible and bounds a disc, say D. It follows that

$$\mathcal{A}(\gamma, x, D) = \mathcal{A}(\alpha, x, D) + \mathcal{A}(\beta, x, \text{constant disc}). \tag{4.6}$$

In order to calculate the second term on the right, observe that the *normalized* Hamiltonian generating the path $\{h_1\phi_s\}$, $s \in [0,t]$ equals

$$F_s \circ h_1^{-1} - \langle F_s \rangle.$$

Since $\phi_s x = x$ and $F_s(x) = 0$, we have that

$$\mathcal{A}(\beta, x, \text{constant disc}) = \int_0^t (F_s \circ h_1^{-1} - \langle F_s \rangle)(h_1\phi_s x)\, ds = a_t,$$

with

$$a_t := -\int_0^t \langle F_s \rangle\, ds.$$

Denote by $\tilde{g}_t \in \mathcal{G}$ the element represented by the path γ. We conclude, with the help of (4.6), that

$$\mathrm{spec}(\tilde{g}_t) = \mathrm{spec}(h) + a_t.$$

Observe now that nondegeneracy of h yields nondegeneracy of $\tilde{g}_t$ for all $t \in [0,1]$. Indeed, by step 1 of the proof, the fixed points of $g_t = h_1\phi_t$ coincide with the ones of h_1, and ϕ_t is the identity near these points. Thus, by the spectrality axiom, $c(\tilde{g}_t) \in \mathrm{spec}(\tilde{g}_t)$, and so $c(\tilde{g}_t) - a_t$ takes values in $\mathrm{spec}(h)$. Moreover, by Exercise 4.3.6, the function $t \mapsto c(\tilde{g}_t) - a_t$ is continuous. Since $\mathrm{spec}(h)$ has empty interior (see [118, Lemma 2.2]), we conclude that this function is constant and equals $c(h)$. Therefore $c(h\phi) = c(h) + a_1$, which proves (4.5) for nondegenerate h.

(3) Assume now that the displacing diffeomorphism h is arbitrary (in general, degenerate). Let H_t be the normalized Hamiltonian generating a path $\{h_t\}$, $t \in [0,1]$, representing h. Fix $\epsilon > 0$. Choose a C^∞-small perturbation H_t' of H_t, giving rise to a nondegenerate element $h' = [\{h_t'\}] \in \mathcal{G}$, such that h_1' displaces a neighborhood of $\bigcup_t \mathrm{supp}(F_t)$. Observe that $h\phi$ and $h'\phi$ are generated by the Hamiltonians $K_t := H_t + F_t \circ h_t^{-1}$ and $K_t' := H_t' + F_t \circ (h_t')^{-1}$, respectively. The perturbation can be chosen so small that $|H_t - H_t'| \le \epsilon$ and $|K_t - K_t'| \le \epsilon$. Thus, by the stability axiom,

$$|c(h) - c(h')| \le \epsilon \quad \text{and} \quad |c(h\phi) - c(h'\phi)| \le \epsilon.$$

By step 2 of the proof,

$$c(h'\phi) = c(h') + a_1,$$

and hence

$$|c(h\phi) - c(h) - a_1| \leq 2\epsilon.$$

Since this inequality holds for every $\epsilon > 0$,

$$c(h\phi) - c(h) - a_1 = 0,$$

which completes the proof. □

Let us indicate the following amusing corollary of Proposition 4.3.5.

Corollary 4.3.7. *Let $U \subset M$ be an open displaceable subset. The Calabi homomorphism on $\widetilde{\mathrm{Ham}}(U)$ descends to $\mathcal{G}(U)$.*

PROOF. Let η and ψ be two elements of $\widetilde{\mathrm{Ham}}(U)$ representing the same element $\phi \in \mathcal{G}(U)$. It follows that $V^{-1} \cdot \mathrm{Cal}(\eta) = V^{-1} \cdot \mathrm{Cal}(\psi) = c(h\phi) - c(h)$, where $h \in \mathcal{G}$ displaces U. □

At the moment we cannot decide whether this statement (which was tacitly assumed in [58]) remains valid for arbitrary (not necessarily displaceable) open U, and whether it is soft or hard. We shall denote by Cal_U the Calabi homomorphism on $\mathcal{G}(U)$.

Theorem 4.3.8. *Let $U \subset M$ be an open displaceable subset, and let $\phi \in \mathcal{G}(U)$.*

(i) $-V^{-1} \cdot \mathrm{Cal}_U(\phi) \leq c(\phi) \leq e_c(U) - V^{-1} \cdot \mathrm{Cal}_U(\phi)$;
(ii) $q(\phi) \leq 2e_c(U)$ *for all* $\phi \in \mathcal{G}(U)$;
(iii) $\sigma(\phi) = -V^{-1} \cdot \mathrm{Cal}_U(\phi)$.

PROOF. Let $h \in \mathcal{G}$ be an element displacing U. By (4.5) and subadditivity,

$$\begin{aligned} -V^{-1} \cdot \mathrm{Cal}_U(\phi) = c(h\phi) - c(h) &\leq c(\phi) \\ &\leq c(h^{-1}) + c(h\phi) = q(h) - V^{-1} \cdot \mathrm{Cal}_U(\phi). \end{aligned}$$

Since h was an arbitrary diffeomorphism displacing U, we get (i).

Adding up equations (i) for ϕ and ϕ^{-1} we get (ii).

Finally, applying (i) to ϕ^n, dividing by n and passing to the limit as $n \to \infty$, we get (iii). □

4.4. Spectral width of a subset

Let us introduce another useful invariant of a subset of M. For an open subset $U \subset M$ define its *spectral width* by

$$w(U) := \sup_{\phi \in \mathcal{G}(U)} q(\phi), \tag{4.7}$$

and for a closed subset $X \subset M$ put

$$w(X) = \inf_{U \supset X} w(U), \tag{4.8}$$

where U runs over all open subsets containing X. By Theorem 4.3.8 and (4.4),

$$w(Y) \leq 2e_c(Y) \leq 4c(\mathbb{1}) + 4e_H(Y), \tag{4.9}$$

where Y is either closed or open. Geometrically, for open U the spectral width $w(U)$ is simply the diameter of the subgroup $\mathcal{G}(U)$ with respect to the spectral pseudo-norm ν_c defined in Section 3.5 above. Note that the spectral width of a subset may be infinite. As we shall see in Corollary 4.6.5 below, the spectral width of any open subset does not vanish.

4.5. Partial symplectic quasi-states

At this point we are able to explain the role of subadditive spectral invariants in function theory on symplectic manifolds. Fix a subadditive spectral invariant $c\colon \mathcal{G} \to \mathbb{R}$. Given a (in general, time-dependent and non-normalized) Hamiltonian F on M, we write ϕ_F^t for the Hamiltonian flow generated by F, and ϕ_F for the element of $\mathcal{G}$ represented by the path $\{\phi_F^t\}$, $t \in [0,1]$. Introduce the *partial symplectic quasi-state* [59] $\zeta\colon C^\infty(M) \to \mathbb{R}$ by

$$\zeta(F) = \sigma(\phi_F) + \langle F \rangle,$$

where $\langle F \rangle$ stands for the mean value of F, see (4.2). Partial symplectic quasi-states possess a number of interesting features which yield a variety of applications to symplectic topology and Hamiltonian dynamics. Let us list some of properties of ζ which immediately follow from earlier results proved in this chapter.

Normalization.

$$\zeta(a) = a \tag{4.10}$$

for any constant function a.

Stability.

$$\min_M (F - G) \leq \zeta(F) - \zeta(G) \leq \max_M (F - G). \tag{4.11}$$

This follows from Proposition 4.3.4.

In particular, ζ is 1-*Lipshitz* in the uniform norm on $C^\infty(M)$ and therefore extends to the space of continuous functions $C(M)$ by $\zeta(F) = \lim \zeta(F_i)$, where the functions F_i are smooth and C^0-converge to a continuous function F. Furthermore, (4.11) yields *monotonicity* of ζ:

$$F \geq G \implies \zeta(F) \geq \zeta(G). \tag{4.12}$$

Homogeneity. The functional ζ is $\mathbb{R}_+$-homogeneous: $\zeta(sF) = s\zeta(F)$ for all $s > 0$.

Invariance. ζ is invariant under the action of $\mathrm{Ham}(M, \omega)$ on functions.

Vanishing. $\zeta(F) = 0$ whenever the support of F is displaceable. This immediately follows from Theorem 4.3.8.

Quasi-subadditivity. $\zeta(F+G) \leq \zeta(F) + \zeta(G)$ whenever $\{F,G\} = 0$. To see this, put $H = F + G$ and denote by f_t, g_t, h_t the Hamiltonian flows generated by F, G, H, respectively. Since F and G Poisson commute, by Exercise 1.3.11 $h_t = f_t g_t$. By the subadditivity, $c(h_t) \leq c(f_t) + c(g_t)$. Dividing by t and passing to the limit as $t \to \infty$, we get that $\zeta(F+G) \leq \zeta(F) + \zeta(G)$.

4.6. The Poisson bracket inequality

The next, less straightforward property of partially symplectic quasi-states plays a crucial role in applications. For $F, G \in C^\infty(M)$ put

$$\Pi(F,G) = |\zeta(F+G) - \zeta(F) - \zeta(G)|.$$

Denote

$$S(F,G) = \min\big(w(\operatorname{supp} F), w(\operatorname{supp} G)\big) \in [0, +\infty].$$

The next proposition is a modification of a result from [66].

Proposition 4.6.1. *For every pair of functions $F, G \in C^\infty(M)$ with $S(F,G) < \infty$,*

$$\Pi(F,G) \leq \sqrt{2S(F,G) \cdot \|\{F,G\}\|}. \tag{4.13}$$

PROOF. Throughout the proof we will denote by ϕ_A the element of $\widetilde{\mathrm{Ham}}(M,\omega)$ generated by a Hamiltonian A.

Assume without loss of generality that F and G have zero mean. Let f_t, g_t be the Hamiltonian flows of F, G, respectively. Set $H = F + G$, $K_t = F + G \circ f_t^{-1}$. Observe that $\phi_K = \phi_F \phi_G$. Calculating

$$G(f_t x) - G(x) = \int_0^t \frac{\mathrm{d}}{\mathrm{d}s} G(f_s x)\, \mathrm{d}s = \int_0^t \{G,F\}(f_s x)\, \mathrm{d}s,$$

we get that $\|H - K_t\| = \|G - G \circ f_t^{-1}\| = \|G \circ f_t - G\| \leq t\|\{F,G\}\|$. Since $\langle K \rangle = \langle F + G \rangle$, we have by Proposition 4.3.4 that

$$|\sigma(\phi_{F+G}) - \sigma(\phi_F \phi_G)| \leq \int_0^1 \|H - K_t\|\, \mathrm{d}t \leq \frac{1}{2}\|\{F,G\}\|.$$

Then

$$\begin{aligned} \Pi(F,G) &= |\sigma(\phi_{F+G}) - \sigma(\phi_F) - \sigma(\phi_G)| \\ &\leq |\sigma(\phi_{F+G}) - \sigma(\phi_F \phi_G)| + |\sigma(\phi_F \phi_G) - \sigma(\phi_F) - \sigma(\phi_G)| \\ &\leq \tfrac{1}{2}\|\{F,G\}\| + I, \end{aligned} \tag{4.14}$$

where $I := |\sigma(\phi_F \phi_G) - \sigma(\phi_F) - \sigma(\phi_G)|$. By Proposition 3.5.3, $I \leq S(F,G)$, and therefore

$$\Pi(F,G) \leq \tfrac{1}{2}\|\{F,G\}\| + S(F,G). \tag{4.15}$$

Let us balance this inequality: Take any $s > 0$ and note that

$$\Pi(sF, sG) = s \cdot \Pi(F,G), \quad \|\{sF, sG\}\| = s^2\|\{F,G\}\|, \quad S(sF, sG) = S(F,G).$$

Substituting into (4.15) and dividing by s we obtain

$$\Pi(F,G) \le \frac{1}{2}s\|\{F,G\}\| + \frac{S(F,G)}{s}.$$

The right-hand side is minimized by $s = \sqrt{2S(F,G)/\|\{F,G\}\|}$, which yields

$$\Pi(F,G) \le \sqrt{2S(F,G)\cdot\|\{F,G\}\|}. \qquad \square$$

Now we are ready to state the following property of partial symplectic quasi-states which complements the quasi-subadditivity presented in Section 4.5:

Partial quasi-additivity. $\zeta(F+G) = \zeta(F)$ provided $\{F,G\} = 0$ and $\operatorname{supp}(G)$ is displaceable. Indeed, $\zeta(G) = 0$ by the vanishing property. Since $S(F,G)$ is finite, the Poisson bracket inequality yields $\zeta(F+G) = \zeta(F)$.

The Poisson bracket inequality yields new information about the spectral pseudo-norm ν_c on $\mathcal{G}$ defined in Section 4.3. Recall that $\nu_c(\mathbb{1}) = 0$ and $\nu_c(\phi) = q(\phi)$ for all $\phi \ne \mathbb{1}$. By Exercise 1.4.3, the center of the group $\mathcal{G}$ coincides with the fundamental group $\mathcal{Z} := \pi_1(\operatorname{Ham}(M), \mathbb{1})$. The next result complements Proposition 3.5.1: the spectral pseudo-norm on $\widetilde{\operatorname{Ham}}$ turns out to be "almost" nondegenerate.

Proposition 4.6.2. *For any $\phi \notin \mathcal{Z}$ we have $\nu_c(\phi) > 0$.*

PROOF. Assume on the contrary that $\phi \notin \mathcal{Z}$ and $\nu_c(\phi) := q(\phi) = 0$. Since the element $\phi \notin \mathcal{Z}$, it displaces some open subset $U \subset M$. By the definition of the spectral displacement energy, $e_c(U) \le q(\phi) = 0$, and hence $e_c(U) = 0$.

Choose Hamiltonian diffeomorphisms $\psi_1, \dots, \psi_N$ so that $\bigcup_{k=1}^N \psi_k(U) = M$ (see Exercise 4.6.3 below). Let $F_1, \dots, F_N$ be a partition of unity subordinated to the cover $\{\psi_k(U)\}$. By the invariance of subadditive spectral invariants, $e_c\big(\psi_k(U)\big) = e_c(U) = 0$, and hence $e_c\big(\operatorname{supp}(F_k)\big) = 0$. Thus, by (4.9), $w\big(\operatorname{supp}(F_k)\big) = 0$. Furthermore, $\zeta(F_k) = 0$ by the vanishing property. Therefore, by (4.13) with $S = 0$

$$\begin{aligned} 1 &= \zeta(F_1 + \dots + F_{N-1} + F_N) \\ &= \zeta(F_1 + \dots + F_{N-1}) + \zeta(F_N) = \dots = \sum_{k=1}^N \zeta(F_k) = 0, \end{aligned}$$

and we get a contradiction. $\square$

Exercise 4.6.3. Let U be a nonempty open subset of a closed symplectic manifold (M,ω). Prove that one can find finitely many $\psi_1, \dots, \psi_N \in \operatorname{Ham}(M,\omega)$ such that $\bigcup_{k=1}^N \psi_k(U) = M$. *Hint*: Recall Exercise 1.3.9.

Exercise 4.6.4. Suppose that $c(\mathbb{1}) = 0$. Prove that the pseudo-norm ν_c is not discrete (see Section 3.5 above for the definition). *Hint*: Look at a sequence of Hamiltonian diffeomorphisms whose supports shrink to a point.

Proposition 4.6.2 readily yields nontriviality of the spectral width associated to any subadditive spectral invariant c on a closed symplectic manifold (M, ω).

Corollary 4.6.5. *The spectral width $w(U)$ of any open subset $U \subset M$ is either positive or infinite. Furthermore, $w(U)$ is finite provided U is displaceable.*

PROOF. Choose any Hamiltonian path $\{f_t\}$, $t \in [0,1]$, $f_0 = \mathbb{1}$ whose Hamiltonian is supported in U and such that $f_1 \neq \mathbb{1}$. Let ϕ be the corresponding element in the universal cover $\widetilde{\mathrm{Ham}}(M, \omega)$. By Proposition 4.6.2, $q(\phi) = \nu_c(\phi) > 0$. Thus, by definition, $w(U) > 0$ or $w(U) = \infty$. Furthermore, if U is displaceable, inequality (4.9) yields finiteness of $w(U)$. □

4.7. Two classes of subadditive spectral invariants

In what follows we deal with two important classes of subadditive spectral invariants. These classes are, in general, not disjoint, nor do they cover all the possibilities.

Class I. The first class consists of subadditive spectral invariants with $c(\mathbb{1}) = 0$. Their importance is due to the fact that, by (4.9), $w(U) \leq 4e_H(U)$. Therefore, the Poisson bracket inequality (4.13) becomes sharper on small scales: the constant $S(F, G)$ in the right-hand side of the inequality is small provided the support of one of the functions has small Hofer displacement energy.

Theorem 4.7.1 ([119, 154]). *Every closed symplectic manifold admits a subadditive spectral invariant with $c(\mathbb{1}) = 0$.*

In Section 12.5 we will prove Theorem 4.7.1 for a certain class of closed symplectic manifolds. Let us present an important corollary of this theorem.

Corollary 4.7.2. *Let (M, ω) be a closed symplectic manifold.*

(i) *Hofer's displacement energy $e_H(U)$ of any open subset $U \subset M$ is either positive or infinite.*
(ii) *Hofer's metric on the group* $\mathrm{Ham}(M, \omega)$ *is nondegenerate.*

Note that this proves Theorem 1.3.15 stated in Chapter 1.

PROOF. Assume that $U \subset M$ is a displaceable subset (otherwise its displacement energy is infinite). Take a subadditive spectral invariant c with $c(\mathbb{1}) = 0$ whose existence is guaranteed by Theorem 4.7.1. By Proposition 4.6.5, the spectral width $w(U)$ is finite and positive. Applying inequality (4.9) we get that

$$e_H(U) \geq \tfrac{1}{4} \cdot w(U) > 0.$$

This proves (i).

Part (ii) of the corollary immediately follows from (i). Indeed, every Hamiltonian diffeomorphism $f \in \mathrm{Ham}(M, \omega)$ with $f \neq \mathbb{1}$ displaces some open subset $U \subset M$. If Hofer's norm of f vanishes, we have that $e_H(U) = 0$, which contradicts (i). $\square$

Class II. The second class consists of subadditive spectral invariants with bounded spectral pseudo-norm ν_c. In this case *every* subset of M has finite spectral width. In particular, setting C to be the diameter of $\mathcal{G}$ with respect to ν_c, we get the following version of the Poisson bracket inequality (4.13):

$$|\zeta(F+G) - \zeta(F) - \zeta(G)| \le \sqrt{2C \cdot \|\{F,G\}\|} \quad \forall F, G \in C^\infty(M). \tag{4.16}$$

As we see, it works for all functions without restrictions on their supports. However, as opposed to the case of Class I subadditive spectral invariants, the inequality does not become sharper on small scales.

As an immediate corollary of (4.16), we get that ζ is linear on all abelian subalgebras of $C^\infty(M)$. Such partial symplectic quasi-states have a number of interesting features which are discussed in detail in the next chapter.

Another consequence of the boundedness of the spectral pseudo-norm is as follows: By Proposition 3.5.1, the function $\sigma\colon \mathcal{G} \to \mathbb{R}$ is a homogeneous quasi-morphism (see Section 4.8 below).

Class II subadditive spectral invariants exist on many interesting symplectic manifolds, such as products of projective spaces [58, 124], blow-ups of arbitrary symplectic manifolds [106, 156], and symplectic toric manifolds [70, 125, 156]. All currently known constructions come from the Floer theory under certain assumptions on the algebraic structure of the quantum homology ring of a symplectic manifold (see Chapter 12 for a detailed discussion).

Let us mention finally that Classes I and II overlap in the following sense: there exist symplectic manifolds admitting a spectral invariant c such that $c(\mathbb{1}) = 0$ and the spectral pseudo-norm ν_c is bounded. For instance, such a c exists on the complex projective space $\mathbb{C}P^n$, see Chapter 12 below.

4.8. Calabi quasi-morphism

Let (M,ω) be a closed symplectic manifold equipped with a Class II subadditive spectral invariant. Let us look more attentively at the corresponding homogeneous quasi-morphism σ. It will be convenient to rescale by putting $\mu := -V \cdot \sigma$, where as above V stands for the volume of M. The following properties of μ follow immediately from Proposition 4.3.4 and Theorem 4.3.8:

(1) **(Calabi property)** For every open displaceable set $U \subset M$ one has $\mu|_{\widetilde{\mathrm{Ham}}(U)} = \mathrm{Cal}_U$.

(2) **(Stability)** Let $f, g \in \widetilde{\mathrm{Ham}}(M,\omega)$ represent paths in $\mathrm{Ham}(M,\omega)$ generated by normalized Hamiltonians F and G, respectively. Then

$$V \cdot \int_0^1 \min_M (F_t - G_t)\, dt \le \mu(g) - \mu(f) \le V \cdot \int_0^1 \max_M (F_t - G_t)\, dt.$$

A homogeneous quasi-morphism with these two properties is called a *Calabi quasi-morphism*. We view this as an abstract definition which makes sense on both closed and open manifolds and which a priori is unrelated to subadditive spectral invariants. Thus we can summarize the above discussion in the following proposition:

Proposition 4.8.1. *Let* (M,ω) *be a closed symplectic manifold and suppose that* c *is a subadditive spectral invariant on* $\widetilde{\mathrm{Ham}}(M,\omega)$ *with bounded spectral pseudo-norm. Then the normalized homogenization* $\mu := -V\cdot\sigma$ *is a Calabi quasi-morphism on* $\widetilde{\mathrm{Ham}}(M,\omega)$.

It is instructive to think about Calabi quasi-morphisms as follows. Let U_α be an open cover of M by displaceable subsets. In contrast to the case of open symplectic manifolds, the group $\mathcal{G}=\widetilde{\mathrm{Ham}}(M,\omega)$ is perfect (that is, it coincides with its commutator subgroup) [13] for any closed symplectic manifold M. Therefore, $\widetilde{\mathrm{Ham}}(M,\omega)$ possesses no nontrivial homomorphisms to $\mathbb{R}$ when M is closed. Nevertheless, for certain closed symplectic manifolds (M,ω), the family of Calabi homomorphisms $\mathrm{Cal}_{U_\alpha}\colon \mathcal{G}(U_\alpha)\to\mathbb{R}$ (which exists by Corollary 4.3.7) extends to a quasi-morphism on $\mathcal{G}$.

The simplest manifolds for which existence of any Calabi quasi-morphism as defined by (i) and (ii) is not known are the standard tori $\mathbb{T}^{2n}=(\mathbb{R}^{2n}/\mathbb{Z}^{2n}, \mathrm{d}p\wedge \mathrm{d}q)$. Interestingly enough, Py [137,138], by using clever geometric methods, discovered homogeneous quasi-morphisms in the case of closed surfaces of genus ≥ 1 which have the Calabi property. It is still not known whether Py's quasi-morphisms are stable. In general, Calabi quasi-morphisms are not unique [27,55,70,125]. They form a convex set whose structure is still far from being understood.

Remark 4.8.2. In some cases, a Calabi quasi-morphism vanishes on $\pi_1(\mathrm{Ham})\subset\widetilde{\mathrm{Ham}}$, and hence descends to a homogeneous quasi-morphism $\overline{\mu}\colon \mathrm{Ham}(M,\omega)\to\mathbb{R}$. This quasi-morphism satisfies the following form of the Calabi property (cf. Remark 4.1.3 above):

(1′) **(Calabi property)** For every open displaceable set $U\subset M$ such that $\omega|_U$ is exact one has $\overline{\mu}|_{\mathrm{Ham}(U)}=\overline{\mathrm{Cal}}_U$.

We refer to [106] for a discussion on this issue.

CHAPTER 5

Symplectic Quasi-states and Quasi-measures

Let (M,ω) be a closed symplectic manifold. Denote by $C(M)$ the space of continuous functions on M. In this chapter we discuss *symplectic quasi-states*, that is, monotone normalized functionals on $C(M)$ which are linear on every Poisson commutative subalgebra, but not necessarily on the whole space. Symplectic quasi-states exist on surfaces as well as on closed symplectic manifolds admitting a subadditive spectral invariant with bounded spectral pseudo-norm. The origins of symplectic quasi-states go back to the foundations of quantum mechanics.

5.1. Symplectic quasi-states

Let (M^{2n},ω) be a closed symplectic manifold. Recall from Section 2.1 that two continuous functions $F,G \in C(M)$ are called Poisson commuting if there exist sequences $F_k, G_k \in C^\infty(M)$ such that

$$F_k \xrightarrow{C^0} F, \quad G_k \xrightarrow{C^0} G, \quad \text{and} \quad \{F_k, G_k\} \xrightarrow{C^0} 0. \tag{5.1}$$

Given $F,G \in C(M)$, we denote by $\mathrm{span}(F,G)$ the subspace of $C(M)$ spanned by F and G.

Definition 5.1.1. A *symplectic quasi-state* on M is a function $\zeta\colon C(M) \to \mathbb{R}$ which satisfies the following axioms:

(1) **(Normalization)** $\zeta(1) = 1$;
(2) **(Positivity)** $F \geq 0 \Rightarrow \zeta(F) \geq 0$;
(3) **(Quasi-linearity)** for any pair of Poisson commuting functions $F,G \in C(M)$, ζ is linear when restricted to $\mathrm{span}(F,G)$.

In this chapter we focus on the case when the symplectic manifold (M,ω) admits a subadditive spectral invariant c with bounded spectral pseudo-norm. Let ζ be the corresponding partial symplectic quasi-state introduced in Section 4.5 above. Recall that we showed that ζ extends to a monotone normalized functional on the space of continuous functions $C(M)$.

Proposition 5.1.2. *Assume that the subadditive spectral invariant c has bounded spectral pseudo-norm. Then the functional $\zeta\colon C(M) \to \mathbb{R}$ is a nonlinear symplectic quasi-state.*

PROOF. We have to check quasi-linearity of ζ. Choose sequences F_k, $G_k \in C^\infty(M)$ as in (5.1). Put $\Pi(F,G) = |\zeta(F+G) - \zeta(F) - \zeta(G)|$. Since Π

is Lipschitz in the C^0-norm, $\Pi(F_k, G_k) \to \Pi(F, G)$. But by Poisson bracket inequality (4.16),

$$|\Pi(F_k, G_k)| \leq \sqrt{2C\|\{F_k, G_k\}\|} \to 0.$$

Therefore,

$$|\zeta(F+G) - \zeta(F) - \zeta(G)| = \Pi(F, G) = 0.$$

This shows that ζ is quasi-linear. In order to check that ζ is nonlinear, take any partition of unity $F_1, \ldots, F_N$ subordinated to an open covering of M by displaceable sets. Non-linearity follows from the fact that $\zeta(F_k) = 0$ by the vanishing property, whereas $\zeta(\sum F_k) = 1$. □

Example 5.1.3 (Median quasi-state). Consider (S^2, σ), the two-sphere with area form σ normalized to have total area 1. For a continuous function $F\colon S^2 \to \mathbb{R}$ denote by Γ_F the set $S^2/\sim$, where $x \sim y$ whenever x and y belong to the same connected component of a level set of F. In other words, Γ_F is obtained from S^2 by collapsing connected components of the level sets of F to points. Denote by $\pi\colon S^2 \to \Gamma_F$ the natural projection.

If F is a smooth Morse function, the set Γ_F has a natural structure of a graph, called *the Reeb graph*. Its vertices correspond to the connected components of level sets of F containing critical points. We call the latter *critical components*. Each such component is either a point of local maximum/minimum of F, or an immersed one-dimensional submanifold whose double points are saddle points of F with the same critical value. By basic Morse theory, the complement to the union of all critical components splits into pair-wise disjoint cylinders diffeomorphic to $S^1 \times \mathbb{R}$. These cylinders correspond to the edges of Γ_F. Take such a cylinder, say E, and look at its boundary $\partial E := \text{Closure}(E) \setminus E$. This set has two connected components corresponding to the "ends" of the cylinder. By definition, the edge defined by E is adjacent to a vertex corresponding to a critical component V if one of the connected components of ∂E is contained in V.

The push-forward $\pi_*(\text{Area}_\sigma)$ of the measure induced by σ gives a probability measure on Γ_F. Note that this is a nonatomic Borel probability measure whose restriction to every open edge is homeomorphic to the Lebesgue measure on an open interval. This construction is illustrated in Figure 5.1 for the height function on the round sphere and on a heart-shaped sphere.

Exercise 5.1.4. Let $F\colon S^2 \to \mathbb{R}$ be a Morse function. Prove that Γ_F is a tree.

Exercise 5.1.5 (Median of a measured tree). Let Γ be any finite tree equipped with a nonatomic Borel probability measure whose restriction to every open edge is homeomorphic to the Lebesgue measure on an open interval. Prove that there exists a unique $m_\Gamma \in \Gamma$, called the *median* of Γ, such that each connected component of $\Gamma \setminus \{m_\Gamma\}$ has measure $\leq \frac{1}{2}$. *Hint*: This exercise is a well-known fact in combinatorial optimization. Its solution can be found in [58, Section 5.3].

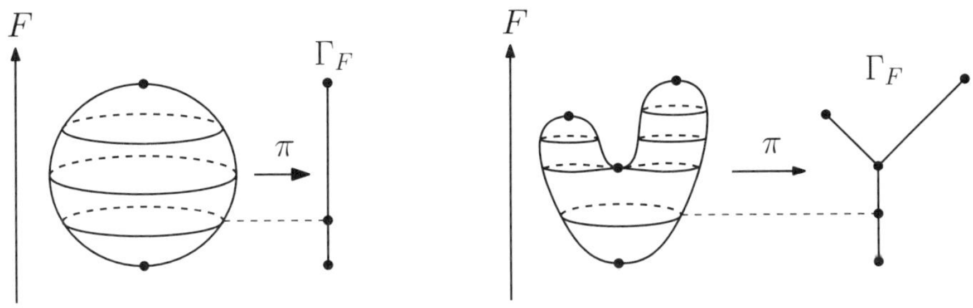

FIGURE 5.1. The Reeb graph

Exercise 5.1.6. Let F be Morse function on S^2 and p be a nonconstant polynomial on $\mathbb{R}$. Prove that connected components of level sets of F and $p(F)$ coincide: $\Gamma_F = \Gamma_{p(F)}$.

Denote by $\mathcal{P} \subset C^\infty(M)$ the subset consisting of all functions of the form $p(F)$, where p is a polynomial and F is a Morse function on S^2. For such a function, put

$$\zeta\big(p(F)\big) := p(F)|_{\gamma_F}, \quad \text{where } \gamma_F := \pi^{-1}(m_{\Gamma_F}). \tag{5.2}$$

We refer to γ_F as *the median level* of F. It follows from Exercise 5.1.6 that ζ is well defined. Indeed, if $p(F) = q(G)$ for a pair of nonconstant real polynomials p, q and a pair of Morse functions F, G, we have that $\Gamma_F = \Gamma_G$ and hence the median levels of F and G coincide: $\gamma_F = \gamma_G$. The latter equality yields $\zeta\big(p(F)\big) = \zeta\big(q(G)\big)$. If one of the polynomials p or q is constant, the statement is trivial.

Exercise 5.1.7. Prove that for any Morse functions $F, G \in C^\infty(S^2)$, $\gamma_F \cap \gamma_G \neq \varnothing$. *Hint*: Use the area condition and the elementary topology of curves on the sphere.

Corollary 5.1.8. *Let p, q be real polynomials and F, G be Morse functions on S^2. If $p(F) \leq q(G)$, then $\zeta\big(p(F)\big) \leq \zeta\big(q(G)\big)$.*

PROOF. By Exercise 5.1.7 there exists $z \in \gamma_F \cap \gamma_G$. Then

$$\zeta\big(p(F)\big) = p\big(F(z)\big) \leq q\big(G(z)\big) = \zeta(G).$$ □

Observe that for every function $H \in \mathcal{P}$ and $c \in \mathbb{R}$ we have that $\zeta(H+c) = \zeta(H) + c$. Together with Corollary 5.1.8 this implies that ζ is Lipschitz in the uniform norm, and hence extends to the space of continuous functions $C(S^2)$.

Proposition 5.1.9. *The functional ζ is a nonlinear symplectic quasi-state on S^2.*

As we shall see, the complete proof of this statement requires some nontrivial work. In order to highlight the difficulty, let us at the moment collect some elementary facts about ζ, which is called *the median quasi-state*.

Clearly, ζ is normalized, that is $\zeta(1) = 1$. Furthermore, by Exercise 5.1.8, ζ is positive: $\zeta(F) \geq 0$ for $F \geq 0$.

Write the sphere as $S^2 = \{x^2 + y^2 + z^2 = 1\} \subset \mathbb{R}^3$. Observe that the median of the height function z on the sphere corresponds to the equator, and hence $\zeta(z^2) = 0$. Similarly, $\zeta(x^2) = \zeta(y^2) = 0$. However, $x^2+y^2+z^2 = 1$, and hence ζ is nonlinear.

In order to finish the proof of Proposition 5.1.9, we need to check quasi-linearity. For Morse functions this is obvious: Indeed, two Morse functions on the sphere Poisson commute if and only if the connected components of their level sets coincide, and hence they have the same median levels. Thus, if F, G and $F + G$ are Morse we are done. The extension to continuous functions is nontrivial. It involves new ideas and will be explained at the end of Section 5.4 below.

The importance of the median symplectic quasi-state in our story is due to the following fact which will be proved later in Section 12.5.

Proposition 5.1.10. *There exists a subadditive spectral invariant with bounded spectral pseudo-norm on the sphere S^2 whose associated partial symplectic quasi-state coincides with the median quasi-state.*

5.2. Quasi-states and the quantum-classical correspondence

The notion of quasi-states originates from John von Neumann's axiomatization of quantum mechanics [159]. Its setting is as follows. Let H be a complex Hilbert space (which we may take to be finite-dimensional). The physical observables are represented by the space $\mathcal{L}(H)$ of all Hermitian operators on H. We consider $\mathcal{L}(H)$ as a real Lie algebra with the bracket given by

$$[A, B]_\hbar = -\frac{\mathrm{i}}{\hbar}[A, B],$$

where $[A, B] = AB - BA$ and $\hbar$ is the Planck constant. The state of a physical system is represented by a functional $\rho\colon \mathcal{L}(H) \to \mathbb{R}$ satisfying the following axioms (the lower index q stands for quantum):

- (1_q) **(Normalization)** $\rho(\mathbb{1}) = 1$.
- (2_q) **(Positivity)** $\rho(A) \geq 0$ if $A \geq 0$.
- (3_q) **(Linearity)** $\rho(aA + bB) = a\rho(A) + b\rho(B)$ for any $a, b \in \mathbb{R}$ and $A, B \in \mathcal{L}(H)$.

Such a functional is called a *quantum state*. According to the statistical interpretation of quantum mechanics, the number $\rho(A)$ represents the expected value of a measurement of A on a system in the state ρ. Von Neumann then proved the following theorem:

Theorem 5.2.1. *For any quantum state ρ there exists nonnegative $U \in \mathcal{L}(H)$ with trace* 1 *such that for any $A \in \mathcal{L}(H)$, $\rho(A) = \mathrm{tr}(UA)$.*

The operator U is called *the density operator* corresponding to ρ. A corollary of this theorem is that for any quantum state there exists an observable $A \in \mathcal{L}(H)$ such that

$$\rho(A^2) - \rho(A)^2 > 0. \tag{5.3}$$

Given the interpretation of $\rho(A)$ as the expectation, the left-hand side of (5.3) represents the variance of the observable A in the state ρ. Thus inequality (5.3) says that there are no dispersion-free states. From this von Neumann concluded that there are no hidden variables in quantum mechanics, i.e., that it is impossible to obtain a deterministic quantum theory by adding further variables describing the state of a system. This conclusion caused a discussion among physicists: it was criticized first by Grete Hermann,[1] whose work on this subject, however, was not properly noticed, and later on by Bohm and Bell (see, e.g., [18]). They argued that the linearity axiom only makes sense for observables A, B that can be measured simultaneously, which in the language of quantum mechanics means that they commute: $[A, B] = 0$. This led to the following definition:

Definition 5.2.2. A quantum quasi-state is a functional $\rho\colon \mathcal{L}(H) \to \mathbb{R}$ which satisfies axioms (1_{q}) and (2_{q}) along with following axiom:

$(3'_{\mathrm{q}})$ **(Quasi-linearity)** $\rho(aA + bB) = a\rho(A) + b\rho(B)$ for any $a, b \in \mathbb{R}$ and any $A, B \in \mathcal{L}(H)$ such that $[A, B] = 0$.

Remark 5.2.3. The condition $[A, B] = 0$ is equivalent to the existence of $C \in \mathcal{L}(H)$ and polynomials $p(t), q(t)$ such that $A = p(C)$ and $B = q(C)$. That is, A and B belong to the subalgebra of $\mathcal{L}(H)$ generated by C.

However, in 1957 Gleason proved the following remarkable theorem:

Theorem 5.2.4 ([75]). *If H has complex dimension 3 or greater, any quantum quasi-state is linear, that is, it is a quantum state.*

Interestingly enough, classical counter-parts of quasi-states are not necessarily linear. In what follows we discuss various aspects of this "anti-Gleason" phenomenon.

5.3. Topological quasi-states

In [1] Aarnes explored potential extensions of Gleason's theorem to classical observables. In his setting the classical phase space is modelled by a compact Hausdorff space X. Observables form the space $C(X)$ of real-valued continuous functions on X, considered as an algebra with respect to the operations of sum and multiplication. A *classical state* is a function $\rho\colon C(X) \to \mathbb{R}$ which satisfies the following axioms (here c stands for classical):

(1_{c}) **(Normalization)** $\rho(1) = 1$.

[1]We refer to Louisa Gilder's lovely book *The age of entanglement: When quantum physics was reborn* for an account of this story.

(2_c) **(Positivity)** $\rho(F) \geq 0$ if $F \geq 0$.
(3_c) **(Linearity)** $\rho(aF + bG) = a\rho(F) + b\rho(G)$ for all $F, G \in C(X)$ and $a, b \in \mathbb{R}$.

Of course, by the Riesz representation theorem, ρ is given by integration with respect to a Borel probability measure on X. Now, let us turn to the notion of quasi-states. Note that in view of Remark 5.2.3, the quasi-linearity of a quantum state can be restated as linearity on singly generated subalgebras of $\mathcal{L}(H)$. Following this analogy, we denote by $\mathcal{B}(F) \subset C(X)$ the *closed* subalgebra generated by $F \in C(X)$, i.e.,

$$\mathcal{B}(F) = \text{Closure}_{C^0}\{p(F) : p \text{ a polynomial}\} = \{f(F) : f \in C(\mathbb{R})\}. \tag{5.4}$$

We call $\mathcal{B}(F)$ a singly generated subalgebra of $C(X)$.

Exercise 5.3.1. Prove the second equality in (5.4). *Hint*: Let p_i be a sequence of polynomials such that $p_i(F)$ converge uniformly to a continuous function G on X. Show that there exists a continuous function g on the image of F with $G = g(F)$. Extend g to a continuous function on the whole $\mathbb{R}$ by using the Tietze extension theorem.

A *topological quasi-state* on X is a functional ρ: $C(X) \to \mathbb{R}$ which satisfies the following axioms:

(1) **(Normalization)** $\rho(1) = 1$.
(2) **(Positivity)** $\rho(F) \geq 0$ if $F \geq 0$.
(3) **(Topological quasi-linearity)** ρ is linear on any singly generated subalgebra of $C(X)$.

Remark 5.3.2. Note that from (1) and (3) it follows that for any $a \in C(X)$ and any $\kappa \in \mathbb{R}$, $\rho(a + \kappa) = \rho(a) + \kappa$.

Example 5.3.3 (Median quasi-state revisited). Let us show that the median quasi-state ζ defined in Example 5.1.3 is a topological quasi-state. We already know that ζ is normalized and monotone. It remains to verify topological quasi-linearity. Take any continuous function $F \in C(M)$ and any pair of continuous functions p, q on $\mathbb{R}$. Choose two sequences of real polynomials p_i, q_i and a sequence of Morse functions F_i on the sphere so that $p_i \circ F_i$ and $q_i \circ F_i$ uniformly converge to $p \circ F$ and $q \circ F$, respectively (we leave the details of this choice as an exercise). Since $p_i \circ F_i + q_i \circ F_i = (p_i + q_i) \circ F_i$, it follows from the definition of ζ on the set $\mathcal{P} \subset C^\infty(M)$ given in (5.2) that

$$\zeta(p_i \circ F_i + q_i \circ F_i) = \zeta(p_i \circ F_i) + \zeta(q_i \circ F_i).$$

Since ζ is continuous in the uniform norm, we conclude that

$$\zeta(p \circ F + q \circ F) = \zeta(p \circ F) + \zeta(q \circ F),$$

which means that ζ is topologically quasi-linear. This completes the proof.

Our next result shows that symplectic quasi-states are special cases of topological quasi-states.

Proposition 5.3.4. *Let (M, ω) be a closed symplectic manifold. Then every symplectic quasi-state on M is a topological quasi-state.*

Exercise 5.3.5. Let $A, B \in C^\infty(M)$ and $a, b \in C^\infty(\mathbb{R})$. Prove that

$$\{a(A), b(B)\} = a'(A)b'(B)\{A, B\}.$$

PROOF OF PROPOSITION 5.3.4. Let ζ be a symplectic quasi-state on M. We must prove that ζ is linear on singly generated subalgebras of $C(M)$. Therefore, let $H \in C(M)$ and let $F, G \in \mathcal{B}(H)$. Then there exist polynomials p_k, q_k such that

$$F = \lim_{k\to\infty} p_k(H), \quad G = \lim_{k\to\infty} q_k(H).$$

Choose a sequence $H_k \in C^\infty(M)$ which converges to H in the C^0-topology. We can find a sequence l_k such that

$$p_k(H_{l_k}) \xrightarrow[k\to\infty]{C^0} F, \quad q_k(H_{l_k}) \xrightarrow[k\to\infty]{C^0} G.$$

But for any k, $\{p_k(H_{l_k}), q_k(H_{l_k})\} \equiv 0$. Thus F and G Poisson commute, and hence for any $a, b \in \mathbb{R}$, $\zeta(aF + bG) = a\zeta(F) + b\zeta(G)$. □

Let us return to the abstract setting of topological quasi-states.

Theorem 5.3.6 (Aarnes [1]). *Let $\rho\colon C(X) \to \mathbb{R}$ be a topological quasi-state. Then ρ is monotone: $a \leq b \Rightarrow \rho(a) \leq \rho(b)$.*

PROOF. In view of Remark 5.3.2, we may assume $0 \leq a \leq b$. Fix $\varepsilon > 0$. Without loss of generality $b \geq a + \varepsilon$, for otherwise we replace b by $b + \varepsilon$ and prove that $\rho(a) \leq \rho(b + \varepsilon) = \rho(b) + \varepsilon$ for all ε. Fix $0 \leq r < \varepsilon$ and choose $N \in \mathbb{N}$ such that $a(X), b(X) \subset [0, Nr]$. For $j = 0, 1, \ldots, N-1$ define $\varphi_j\colon \mathbb{R} \to [0, 1]$ such that

- $\varphi_j(s) = 0$ for all $s \leq jr$.
- $\varphi_j(s) = r$ for all $s \geq (j+1)r$.
- φ_j increases linearly with slope 1 on $[jr, (j+1)r]$.

(see Figure 5.2).

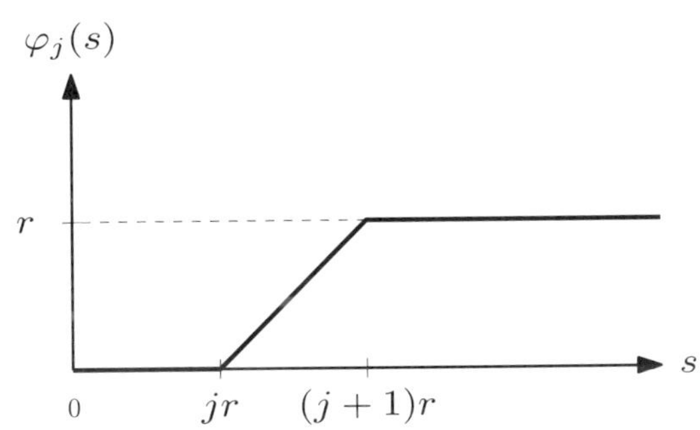

FIGURE 5.2. The function φ_j

Exercise 5.3.7. Prove that for any $s \in [0, Nr]$,

$$\varphi_0(s) + \cdots + \varphi_{N-1}(s) = s.$$

Therefore, $a = \sum_j \varphi_j(a)$, $b = \sum_j \varphi_j(b)$, and hence

$$\rho(a) = \sum_j \rho\big(\varphi_j(a)\big), \quad \rho(b) = \sum_j \rho\big(\varphi_j(b)\big). \tag{5.5}$$

Fix j, and set $f = \varphi_j(a)$, $g = \varphi_j(b)$. Then since φ_j is monotone, $f \leq g$. Let $x \in X$ such that $f(x) > 0$. This implies that $a(x) > jr$ and hence

$$b(x) \geq a(x) + \varepsilon > jr + \varepsilon > (j+1)r.$$

Therefore, $g(x) = r$. Set $h = f + g - r$, and note that $h(x) = f(x) > 0$ whenever $f(x) > 0$, and $h(x) = g(x) - r \leq 0$ whenever $f(x) = 0$. Therefore,

$$f(x) = \max\{0, h(x)\}, \quad g(x) - r = \min\{0, h(x)\}.$$

Therefore, setting

$$\alpha(s) = \max\{0, s\}, \quad \beta(s) = \min\{0, s\} + r,$$

we note that α, β are continuous, and $f = \alpha(h)$, $g = \beta(h)$. In particular, $f, g \in \mathcal{B}(h) \subset C(X)$, and thus

$$\rho\big(\varphi_j(b)\big) = \rho(g) = \rho(f) + \rho(g - f) \geq \rho(f) = \rho\big(\varphi_j(a)\big).$$

Since this holds for any $0 \leq j \leq N-1$, by (5.5), $\rho(a) \leq \rho(b)$. □

5.4. Quasi-measures

In this section we prove an analogue of the Riesz representation theorem for topological quasi-states discovered by Aarnes [1]: it turns out quasi-states are represented by set functions called *quasi-measures.*[2]

Let X be a compact Hausdorff space. Denote by $\mathrm{CO}(X)$ the collection of all subsets of X that are either closed or open.

Definition 5.4.1. A *quasi-measure* on X is a function $\tau\colon \mathrm{CO}(X) \to [0,1]$ which satisfies the following axioms:

(1) **(Normalization)** $\tau(X) = 1$.
(2) **(Monotonicity)** $A \subset B \Rightarrow \tau(A) \leq \tau(B)$.
(3) **(Additivity)** $\tau(A \cup B) = \tau(A) + \tau(B)$ for any $A, B \in \mathrm{CO}(X)$ such that $A \cup B \in \mathrm{CO}(X)$ and $A \cap B = \varnothing$.
(4) **(Regularity)** If $U \subset X$ is open then

$$\tau(U) = \sup\{\tau(K) : K \subset U,\ K \text{ closed}\}.$$

Remark 5.4.2. Observe that the regularity and additivity axioms imply that for a closed subset $K \subset X$ one has

$$\tau(K) = \inf\{\tau(U) : U \supset K,\ U \text{ open}\}.$$

[2]Quasi-measures are also called topological measures.

Exercise 5.4.3. Let $A_1, \ldots, A_n \in \mathrm{CO}(X)$ be pairwise disjoint sets such that $\bigcup_k A_k \in \mathrm{CO}(X)$. Prove that $\tau(\bigcup_k A_k) = \sum_k \tau(A_k)$. *Hint*: Note that $A_{k_1} \cup \cdots \cup A_{k_l}$ need not be in $\mathrm{CO}(X)$ in general. Find collections of subsets $A_{k_1}, \ldots, A_{k_l}$ for which the unions are in $\mathrm{CO}(X)$.

Before presenting specific examples of quasi-measures, let us mention that quasi-measures on closed d-dimensional manifolds are completely determined by their values on d-dimensional compact connected submanifolds with boundary. This was pointed out by Zapolsky [165]:

Exercise 5.4.4. Let M be a closed manifold. Assume that a compact K is contained in an open subset $U \subset M$. Prove that there exists a compact submanifold with boundary W with $K \subset W \subset U$. Conclude that $\tau(K) = \inf \tau(W)$, where the infimum is taken over all submanifolds with boundary W containing K.

Exercise 5.4.5. Deduce from the previous exercise that every quasi-measure on the sphere S^2 is determined by its values on closed embedded discs.

Example 5.4.6. Consider the following quasi-measure on S^2. Pick three points $a, b, c \in S^2$. For a closed disc D set

$$\tau(D) = \begin{cases} 1, & \text{if } D \text{ contains at least two of the points,} \\ 0, & \text{otherwise.} \end{cases}$$

For instance, take a, c to be the north and south poles, respectively, and b any point on the equator. Take P, R to be closed caps of area $< \frac{1}{2}$ around a, c, respectively, and $Q = S^2 \setminus (P \cup R)$. Then by definition, $\tau(P) = \tau(R) = 0$, and hence $\tau(Q) = 1$. Observe that although Q contains only one of the points a, b, c, it has full measure. This does not conflict with our definition, since Q is not a topological disc. Thus the quasi-measure of a subset is sensitive to its topology.

Example 5.4.7. Again on S^2 consider the quasi-measure whose value on a closed disc D is

$$\tau(D) = \begin{cases} 1, & \text{if } \mathrm{Area}(D) \geq \frac{1}{2}, \\ 0, & \text{otherwise.} \end{cases}$$

Let $D^0_\pm$ be the open top and bottom hemispheres, respectively and let $D_\pm$ be their closure. Then $\tau(D_\pm) = 1$ and hence $\tau(D^0_\pm) = 0$. Let $L \subset S^2$ be the equator. Since $S^2 = D^0_+ \cup D^0_- \cup L$, $\tau(L) = 1$. Now take P, Q to be closed discs of area $< \frac{1}{2}$ whose union covers L. Then $\tau(P) = \tau(Q) = 0$ but $\tau(P \cup Q) = 1$. Therefore, τ is not subadditive.

We now turn to the correspondence between quasi-states and quasi-measures. To this end, we introduce integration with respect to quasi-measures. Let τ be a quasi-measure on X. Let $a \in C(X)$. For $\alpha \in \mathbb{R}$

set

$$K_\alpha^a = \{x \in X : a(x) \geq \alpha\}.$$

Define

$$\hat{a} \colon \mathbb{R} \to [0,1], \quad \hat{a}(\alpha) = \tau(K_\alpha^a).$$

Obviously, $\hat{a}$ is a nonincreasing function.

Exercise 5.4.8. Prove that $\hat{a}$ is left-continuous.

Define a Borel measure on $\mathbb{R}$ by setting

$$\sigma_a\big([\alpha,\beta)\big) = \hat{a}(\alpha) - \hat{a}(\beta).$$

Definition 5.4.9. Define the integral of a with respect to τ by

$$\int_X a\,\mathrm{d}\tau = \int_{\mathbb{R}} \alpha\,\mathrm{d}\sigma_a(\alpha).$$

We will occasionally write $\int$ for $\int_X$ when the space X is clear from the context. Note that this is the standard Riemann–Stieltjes integral of the function $-\hat{a}$, and thus we can give an explicit formula for the integral:

$$\int_X a\,\mathrm{d}\tau = \max a - \int_{\min a}^{\max a} \tau(\{a \leq \alpha\})\,\mathrm{d}\alpha.$$

Example 5.4.10. Consider the quasi-measure on S^2 defined in Example 5.4.6. Take $z_1 < z_2 < z_3$ and take $A_k \in S^2$ with z-coordinate z_k. Take $a \in C(S^2)$ to be the height function. Then

$$\tau(K_\alpha^a) = \begin{cases} 1, & \text{if } \alpha \leq z_2, \\ 0, & \text{if } \alpha > z_2. \end{cases}$$

That is, $\sigma_a = \delta_{z_2}$, the Dirac delta measure concentrated at z_2.

Theorem 5.4.11. *The map*

$$C(X) \to \mathbb{R}, \quad a \mapsto \int a\,\mathrm{d}\tau$$

is a topological quasi-state.

The proof of Theorem 5.4.11 is quite long. We therefore break it into several steps. Our first step is to prove that any quasi-measure on a compact interval is a measure. In fact, following Wheeler [161], Shakhmatov [146], Grubb and LaBerge [79], we will prove a slightly stronger result, for which we need the following definition: Let X be a compact Hausdorff space. Let $\mathscr{F}$ be any family of subsets of X. We say that $\mathscr{F}$ has order $\leq n$ if any $n+2$ sets in $\mathscr{F}$ have empty intersection. We say that $\operatorname{CovDim} X \leq n$ if any finite open cover of X has a refinement of order $\leq n$, and define the *covering dimension* of X as

$$\operatorname{CovDim} X = \min\{n : \operatorname{CovDim} X \leq n\}.$$

Example 5.4.12. It is easy to see that the unit interval $[0,1]$ has covering dimension 1. Indeed, given any finite cover of $[0,1]$ we can, by possibly taking a refinement, assume that it is composed of open intervals. Pick an interval containing 0 with maximal right endpoint $x \in (0,1]$. Now pick an interval containing x with maximal right endpoint. Proceeding by induction we get a refinement of order ≤ 1, hence $\operatorname{CovDim}[0,1] \leq 1$. Since $[0,1]$ is connected, its covering dimension cannot be less than 1. Thus $\operatorname{CovDim}[0,1] = 1$.

We will prove that if X has covering dimension at most 1, any quasi-measure on X is subadditive. For the rest of this section, we use the notation $\overline{Y}$ for the complement of a subset $Y \subset X$, that is $\overline{Y} = X \setminus Y$. We start with the following lemma:

Lemma 5.4.13. *Let ν be a quasi-measure on X. Let $\{U_1, U_2, U_3\}$ be an open cover of X such that $U_1 \cap U_2 \cap U_3 = \varnothing$. Then $\nu(U_1)+\nu(U_2)+\nu(U_3) \geq 1$.*

Proof. Represent U_1 as the disjoint union of three sets from $\mathrm{CO}(X)$

$$U_1 = (\overline{U_2 \cup U_3}) \sqcup (U_1 \cap U_2) \sqcup (U_1 \cap U_3).$$

We have similar decompositions for U_2, U_3, which give:

$$\begin{aligned}
\nu(U_1) &= \nu(\overline{U_2 \cup U_3}) + \nu(U_1 \cap U_2) + \nu(U_1 \cap U_3)\\
\nu(U_2) &= \nu(\overline{U_1 \cup U_3}) + \nu(U_1 \cap U_2) + \nu(U_2 \cap U_3)\\
\nu(U_3) &= \nu(\overline{U_1 \cup U_2}) + \nu(U_1 \cap U_3) + \nu(U_2 \cap U_3)
\end{aligned}$$

Adding up, we obtain

$$\begin{aligned}
\nu(U_1) + \nu(U_2) + \nu(U_3) \geq \nu(\overline{U_2 \cup U_3}) &+ \nu(\overline{U_1 \cup U_3}) + \nu(\overline{U_1 \cup U_2})\\
&+ \nu(U_1 \cap U_2) + \nu(U_1 \cap U_3) + \nu(U_2 \cap U_3) = 1,
\end{aligned}$$

where the last equality holds since the sets on the right-hand side form a disjoint cover of X by sets from $\mathrm{CO}(X)$. □

Lemma 5.4.14. *Assume* $\operatorname{CovDim} X \leq 1$. *Then every quasi-measure on X is subadditive.*

Proof. Let ν be a quasi-measure on X. Assume, on the contrary, that ν is not subadditive, and let $U_1, U_2 \subset X$ be open sets such that $\nu(U_1 \cup U_2) > \nu(U_1) + \nu(U_2)$. Then

$$\nu(U_1) + \nu(U_2) + \nu(\overline{U_1 \cup U_2}) = \nu(U_1) + \nu(U_2) + 1 - \nu(U_1 \cup U_2) < 1.$$

By Remark 5.4.2, for every $\delta > 0$ there exists an open U_3 such that

$$\overline{U_1 \cup U_2} \subset U_3, \quad \nu(U_3) \leq \nu(\overline{U_1 \cup U_2}) + \delta.$$

Choosing δ small enough, we get

$$\nu(U_1) + \nu(U_2) + \nu(U_3) < 1.$$

Now, U_1, U_2, U_3 form an open cover of X, and since the covering dimension of X is ≤ 1, it has a refinement $\{W_i\}_{i\in I}$ so that every triple of W_i's do not intersect. Take any function $\alpha\colon I \to \{1,2,3\}$ such that $W_i \subset U_{\alpha(i)}$, and put

$$V_j = \bigcup_{i\in\alpha^{-1}(j)} W_i, \quad j \in \{1,2,3\}.$$

It follows that V_1, V_2, V_3 is a refinement of U_1, U_2, U_3 such that $V_1\cap V_2\cap V_3 = \varnothing$. But, since $V_k \subset U_k$,

$$\nu(V_1) + \nu(V_2) + \nu(V_3) < 1,$$

contradicting Lemma 5.4.13. Therefore, ν is subadditive. □

In particular, every quasi-measure on $[0,1]$ is subadditive. Next, we want to deduce that such a quasi-measure is in fact (the restriction of) a measure. This will be a consequence of the following result:

Theorem 5.4.15. *Every subadditive quasi-measure ν extends to a measure. Moreover, the integral with respect to ν is the standard Lebesgue integral.*

For the proof of this theorem we need several auxiliary statements and constructions. Let ν be a subadditive quasi-measure on X. For $E \subset X$, set

$$\nu^*(E) = \inf\{\nu(U) : U \supset E,\, U \text{ open}\}.$$

It is easy to see that by the Remark 5.4.2, for $E \in \mathrm{CO}(X)$, $\nu^*(E) = \nu(E)$.
The subadditivity of ν extends to ν^*:

Proposition 5.4.16. *ν^* is subadditive.*

PROOF. Let $E_1, E_2 \subset X$, and let $U_k \supset E_k$ such that $\nu(U_k) \leq \nu^*(E_k)+\varepsilon$. In particular, $E_1 \cup E_2 \subset U_1 \cup U_2$, and hence

$$\nu^*(E_1 \cup E_2) \leq \nu(U_1 \cup U_2) \leq \nu(U_1) + \nu(U_2) \leq \nu^*(E_1) + \nu^*(E_2) + 2\varepsilon.$$

Since $\varepsilon > 0$ is arbitrary, the result follows. □

Define

$$\mathscr{A} = \left\{ E \subset X : \begin{array}{l}\forall\varepsilon > 0\ \exists \text{ open } U \supset E \text{ and closed } F \subset E\\ \text{so that } \nu(U) - \nu(F) < \varepsilon\end{array}\right\}.$$

It is easily verified that $\mathscr{A}$ is an algebra, i.e., is closed under intersection, union, and taking complement. Note also that $\mathscr{A}$ contains the Borel algebra. Next we verify the additivity of ν^* on $\mathcal{A}$.

Lemma 5.4.17. *For every pair of subsets $E_1, E_2 \in \mathscr{A}$ with $E_1\cap E_2 = \varnothing$ one has*

$$\nu^*(E_1 \cup E_2) = \nu^*(E_1) + \nu^*(E_2).$$

PROOF. Due to the subadditivity of ν^*, it suffices to prove that

$$\nu^*(E_1 \cup E_2) \geq \nu^*(E_1) + \nu^*(E_2).$$

Fix $\varepsilon > 0$, and let $F_1 \subset E_1$ and $F_2 \subset E_2$ be closed sets with

$$\nu(F_i) \geq \nu^*(E_i) - \varepsilon, \quad i = 1, 2.$$

None that $F_1 \cap F_2 = \varnothing$ and hence $\nu(F_1 \cup F_2) = \nu(F_1) + \nu(F_2)$. Finally, $E_1 \cup E_2 \supset F_1 \cup F_2$ and hence

$$\nu^*(E_1 \cup E_2) \geq \nu(F_1 \cup F_2) \geq \nu^*(E_1) + \nu^*(E_2) - 2\varepsilon.$$

Since ε was arbitrary, the result follows. □

Finally, we check that ν^* is σ-additive:

Theorem 5.4.18 (Grubb–LaBerge [78]**).** *ν^* is σ-additive on $\mathscr{A}$, and hence extends to a measure on the σ-algebra generated by $\mathscr{A}$. In particular, ν^* extends to a Borel measure.*

PROOF. Let $A_k \in \mathscr{A}$ be such that $A_k \cap A_j = \varnothing$ for $k \neq j$ and $A := \bigcup_{k=1}^{\infty} A_k \in \mathscr{A}$. First let us show that

$$\nu^*(A) \leq \sum_{i=1}^{\infty} \nu^*(A_k).$$

Indeed, let $\varepsilon > 0$ and pick open $U_k \supset A_k$ such that $\nu^*(A_k) \geq \nu^*(U_k) - \varepsilon/2^{k+1}$. Denote $C = \bigcup_{k=1}^{\infty} U_k$. Since C is open, by regularity we can find a closed $F \subset C$ with

$$\nu(C) \leq \nu(F) + \frac{\varepsilon}{2}. \tag{5.6}$$

Since $F \subset X$ is closed, it is compact, and the U_k form an open cover of F, hence $F \subset U_{k_1} \cup \ldots \cup U_{k_n}$. Therefore, using the finite subadditivity of ν^*, we get

$$\nu^*(F) \leq \sum_{j=1}^{n} \nu^*(U_{k_j}) \leq \sum_{k=1}^{\infty} \left(\nu^*(A_k) + \frac{\varepsilon}{2^{k+1}} \right) = \sum_{k=1}^{\infty} \nu^*(A_k) + \frac{\varepsilon}{2}.$$

Combining this with (5.6), we get

$$\nu(C) \leq \sum_{k=1}^{\infty} \nu^*(A_k) + \varepsilon.$$

Note that $\nu^*(A) \leq \nu(C)$ by the definition of ν^*. Since $\varepsilon > 0$ is arbitrary, the last two inequalities yield

$$\nu^*(A) \leq \sum_{k=1}^{\infty} \nu^*(A_k).$$

For the converse direction, note that the finite additivity and monotonicity of ν^* on $\mathscr{A}$ give

$$\sum_{k=1}^{n} \nu^*(A_k) = \nu^*\left(\bigcup_{k=1}^{n} A_k\right) \leq \nu^*(A).$$

Therefore, the increasing sequence $\sum_{k=1}^{n} \nu^*(A_k)$ is bounded from above, and hence converges. Thus

$$\sum_{k=1}^{\infty} \nu^*(A_k) \leq \nu^*(A). \qquad \square$$

Exercise 5.4.19. Show that when ν extends to a regular Borel measure, our definition of $\int f \, \mathrm{d}\nu$ coincides with the usual Lebesgue integral.

Theorem 5.4.18 and Exercise 5.4.19 combined yield the proof of Theorem 5.4.15.

Our next step is to define the push-forward of a quasi-measure:

Exercise 5.4.20. Let $\varphi\colon X \to Y$ be a continuous map and let τ be a quasi-measure on X. Define its push-forward as

$$\varphi_*\tau\colon \mathrm{CO}(Y) \to \mathbb{R}, \quad \varphi_*\tau(A) = \tau(\varphi^{-1}A).$$

Prove that $\varphi_*\tau$ is a quasi-measure on Y and for any $a \in C(Y)$,

$$\int_Y a \, \mathrm{d}\varphi_*\tau = \int_X \varphi^* a \, \mathrm{d}\tau. \tag{5.7}$$

After this preparatory work, we can now prove Theorem 5.4.11:

PROOF OF THEOREM 5.4.11. Let τ be a quasi-measure on X. We must show that the map $a \mapsto \int a \, \mathrm{d}\tau$ satisfies the axioms of a topological quasi-state. The normalization and positivity axioms are obvious, so it remains to verify the topological quasi-linearity. So, let $a, b \in \mathcal{B}(c) \subset C(X)$ for some $c \in C(X)$. We must show that

$$\int (a+b) \, \mathrm{d}\tau = \int a \, \mathrm{d}\tau + \int b \, \mathrm{d}\tau.$$

By the definition of a singly generated subalgebra, there exist $\varphi, \psi \in C(\mathbb{R})$ such that $a = \varphi(c)$, $b = \psi(c)$. Let $J \subset \mathbb{R}$ be a compact interval such that $c(X) \subset J$. Consider the push-forward quasi-measure $c_*\tau$ on J. Any closed interval has covering dimension 1, and so by Lemma 5.4.14 and Theorem 5.4.15, $c_*\tau$ extends to a measure and integration with respect to $c_*\tau$ is the usual Lebesgue integral. Therefore, using (5.7) we get

$$\int_X (a+b) \, \mathrm{d}\tau = \int_J (\varphi+\psi) \, \mathrm{d}c_*\tau = \int_J \varphi \, \mathrm{d}c_*\tau + \int_J \psi \, \mathrm{d}c_*\tau = \int_X a \, \mathrm{d}\tau + \int_X b \, \mathrm{d}\tau,$$

and the proof is complete. $\square$

This completes the proof of Theorem 5.4.11, and establishes one direction in the correspondence between topological quasi-states and quasi-measures. Aarnes [1] proved that in fact this correspondence defines a bijection between quasi-measures and topological quasi-states. Let $\rho\colon C(X) \to \mathbb{R}$ be a topological quasi-state. Define, for a closed subset $A \subset X$

$$\tau(A) = \inf\{\rho(a) : a\colon X \to [0,1] \text{ continuous, } a|_A \equiv 1\}. \tag{5.8}$$

Heuristically speaking, $\tau(A)$ is the value taken by ρ on the (discontinuous) indicator function χ_A. For an open subset $A \subset X$ we set $\tau(A) = 1 - \tau(X \setminus A)$. Aarnes [1] proved the following

Theorem 5.4.21 (Aarnes). *The set function τ is a quasi-measure and*

$$\int a\, \mathrm{d}\tau = \rho(a) \quad \forall a \in C(X).$$

PROOF. We leave it to the reader to check that τ is a quasi-measure. For the rest we essentially follow Aarnes' argument [1, p. 56]: By Theorem 5.4.11, the functional

$$\xi\colon C(X) \to \mathbb{R}, \quad a \mapsto \int a\, \mathrm{d}\tau$$

is a topological quasi-state. In order to prove that $\xi = \rho$, fix $a \in C(X)$. Let $[p, q] \subset \mathbb{R}$ be an interval containing the image of a. Observe that the functional sending a function $f \in C([p,q])$ to $\rho(f \circ a)$ is normalized, positive and linear, and hence it is given by integration against some Borel probability measure μ on $[p, q]$. It now suffices to show that $\mu = a_*\tau$.

Take any $\alpha \in [p, q]$. Let $\alpha_n := \alpha - 1/n$. Put $Y_n = [\alpha_n, q]$ and $Y = [\alpha, q]$. Let u_n and u be the indicator functions of the sets Y_n and Y, respectively. Let v_n be a continuous function on $[p, q]$ with

$$u \le v_n \le u_n. \tag{5.9}$$

First, observe that by (5.8) and monotonicity of ρ,

$$a_*\tau(Y) \le \rho(a^* v_n) \le a_*\tau(Y_n).$$

By Exercise 5.4.8, $a_*\tau(Y_n) \to a_*\tau(Y)$ as $n \to \infty$, and thus $\rho(a^*v_n) \to a_*\tau(Y)$.

Second, since μ is a Borel measure and $\rho(a^*v_n) = \int v_n\, \mathrm{d}\mu$, we get from (5.9) that $\rho(a^*v_n) \to \mu(Y)$. Therefore $\mu(Y) = a_*\tau(Y)$. That is, $\mu([\alpha, q]) = a_*\tau([\alpha, q])$ for every α. It follows that $\mu = a_*\tau$, as required. □

Let us state the following consequence of Theorem 5.4.21 which will be used later on in the book.

Proposition 5.4.22. *Let $\zeta\colon C(X) \to \mathbb{R}$ be a topological quasi-state and let τ be the associated quasi-measure. Let $Y \subset X$ be a closed subset. Then $\tau(Y) = 1$ if and only if for any function $F \in C(X)$ with $F|_Y \equiv c$ one has $\zeta(F) = c$.*

PROOF. The "if" direction immediately follows from the definition of τ given in (5.8). Let us prove the "only if" statement. By Theorem 5.4.21, it suffices to show that

$$\int_X F \, \mathrm{d}\tau = c.$$

Indeed, for any $\alpha > c$ one has $\{F \geq \alpha\} \subset X \setminus Y$ and for any $\alpha \leq c$ one has $\{F \geq \alpha\} \supset Y$. Therefore

$$\tau(\{F \geq \alpha\}) = \begin{cases} 1, & \text{if } \alpha \leq c, \\ 0, & \text{if } \alpha > c. \end{cases}$$

It follows that the measure σ_F is the Dirac measure at c, $\sigma_F = \delta_c$. Therefore,

$$\zeta(F) = \int_X F \, \mathrm{d}\tau = \int_{\mathbb{R}} \alpha \, \mathrm{d}\sigma_F(\alpha) = c. \qquad \square$$

Next, we return to the connection between symplectic and topological quasi-states. We saw in the previous section that every symplectic quasi-state is a topological quasi-state. The converse is not true in general, but it is true in the two-dimensional case:

Theorem 5.4.23. *Suppose the symplectic manifold (M, ω) is a closed oriented surface. Then every topological quasi-state is a symplectic quasi-state.*

We prove this in several steps. For the rest of this section, M stands for a closed oriented surface.

Since by Theorem 5.3.6 every topological quasi-state is monotone, it remains to verify the quasi-linearity axiom. We begin by proving quasi-linearity on smooth functions. We will use a version of the Sard lemma presented in [50].

Lemma 5.4.24. *Let $\Phi\colon M \to \mathbb{R}^2$ be a smooth map. Then the set of critical values of Φ in $\mathbb{R}^2$ has covering dimension ≤ 1.*

By using this lemma, we can prove the following statement.

Lemma 5.4.25. *Let $\rho\colon C(M) \to \mathbb{R}$ be a topological quasi-state, and let $F, G \in C^\infty(M)$ be Poisson commuting functions. Then $\rho|_{\mathrm{span}(F,G)}$ is linear.*

PROOF. Set

$$\Phi\colon M \to \mathbb{R}^2, \quad x \mapsto \big(F(x), G(x)\big)$$

and note that $\det \Phi_* = \{F, G\} \equiv 0$. Therefore, the image Δ of Φ consists only of critical values, and hence by Lemma 5.4.24 has covering dimension at most 1. Let τ be the quasi-measure associated with ρ. Then $\Phi_*\tau$ is a quasi-measure on Δ which, by Lemma 5.4.14 and Theorem 5.4.15, extends to a measure. Denote by s_1, s_2 the coordinate functions on $\mathbb{R}^2$ and observe that $F = \Phi^* s_1$ and $G = \Phi^* s_2$. Therefore,

$$\begin{aligned}\rho(F+G) &= \int_M (F+G)\,\mathrm{d}\tau = \int_\Delta (s_1+s_2)\,\mathrm{d}\Phi_*\tau \\ &= \int_\Delta s_1\,\mathrm{d}\Phi_*\tau + \int_\Delta s_2\,\mathrm{d}\Phi_*\tau \\ &= \int_M F\,\mathrm{d}\tau + \int_M G\,\mathrm{d}\tau = \rho(F)+\rho(G). \qquad \Box\end{aligned}$$

In order to extend the quasi-linearity to continuous functions, we need the following lemma [164]:

Lemma 5.4.26. *There exists a constant $C > 0$ such that the following holds. For any $F, G \in C^\infty(M)$ such that $\|\{F,G\}\| \leq \varepsilon$ there exist $F', G' \in C^\infty(M)$ such that*

$$\|F - F'\| \leq C\sqrt{\varepsilon}, \quad \|G - G'\| \leq C\sqrt{\varepsilon}, \quad \textit{and} \quad \{F', G'\} \equiv 0.$$

The proof is technical and we shall omit it, but the idea behind it is simple: consider again the map $\Phi(x) = \big(F(x), G(x)\big)$. Since $\|\Phi_*\| = \|\{F,G\}\| \leq \varepsilon$, its image will have small area, of order ε. One then approximates it by a one-dimensional set, which is the image of another map $\Phi'(x) = \big(F'(x), G'(x)\big)$, with approximation error $\sim \sqrt{\varepsilon}$.

Proof of Theorem 5.4.23. Let $F, G \in C(M)$ be Poisson commuting functions. Choose sequences $F_k, G_k \in C^\infty(M)$ such that

$$F_k \xrightarrow{C^0} F, \quad G_k \xrightarrow{C^0} G, \quad \{F_k, G_k\} \xrightarrow{C^0} 0.$$

By Lemma 5.4.26, we can choose new sequences $F'_k, G'_k \in C^\infty(M)$ such that

$$F'_k \xrightarrow{C^0} F, \quad G'_k \xrightarrow{C^0} G, \quad \{F'_k, G'_k\} \equiv 0.$$

By Lemma 5.4.25, $\rho(F'_k + G'_k) = \rho(F'_k) + \rho(G'_k)$. Taking the limit $k \to \infty$, the result follows. $\Box$

Remark 5.4.27 (Zapolsky). The appeal to Lemma 5.4.24 is actually unnecessary. From the proof of Lemma 5.4.26 in [164] one can in fact extract that for the approximating commuting Hamiltonians F'_k and G'_k, the image of the map $(F'_k, G'_k)\colon M \to \mathbb{R}^2$ is a finite union of intervals, and hence has covering dimension at most 1. Therefore, we get that ζ is linear on $\mathrm{span}(F'_k, G'_k)$ without requiring Lemma 5.4.24.

Now we are ready to prove Proposition 5.1.9 claiming that the median quasi-state ζ on S^2 is indeed a symplectic quasi-state. We have seen in Example 5.3.3 that the median quasi-state ζ on S^2 is a topological quasi-state. Theorem 5.4.23 yields the claim.

Exercise 5.4.28. Prove that the quasi-measure τ associated to the median quasi-state on S^2 vanishes on all closed embedded discs of area $< \frac{1}{2}$ (and hence automatically equals 1 on all closed embedded discs of area $\geq \frac{1}{2}$). Thus τ is the quasi-measure considered in Example 5.4.7.

Exercise 5.4.29. Prove that the median quasi-state is the unique symplectic quasi-state on S^2 satisfying the vanishing property from the previous exercise. *Hint*: Prove that for any other such symplectic quasi-state the associated quasi-measure coincides with the quasi-measure associated with the median quasi-state (use Exercises 5.4.5 and 5.4.28). Deduce, using Theorem 5.4.21, that the quasi-states coincide as well.

5.5. Reduction of symplectic quasi-states

In this section we describe, following Borman [26, 27], the behavior of symplectic quasi-states under Marsden–Weinstein reduction (see Section 1.7). Recall that, by Proposition 5.1.2, if (M, ω) admits a subadditive spectral invariant with bounded spectral quasi-norm the associated partial quasi-state ζ extends to a genuine symplectic quasi-state $\zeta\colon C(M) \to \mathbb{R}$. Moreover, symplectic quasi-states arising in this manner enjoy the additional properties of *vanishing*: $\zeta(F) = 0$ whenever $\operatorname{supp} F$ is displaceable and Ham-*invariance*: $\zeta(F \circ \phi) = \zeta(F)$ for all $F \in C(M)$ and $\phi \in \mathrm{Ham}$, and satisfy the *Poisson bracket inequality*:

$$|\zeta(F+G) - \zeta(F) - \zeta(G)| \le C\sqrt{\|\{F, G\}\|}.$$

As we will see, these three properties are preserved by the reduction procedure.

We begin by describing the class of hypersurfaces we will focus on.

Definition 5.5.1. A compact hypersurface $\Sigma \subset (M, \omega)$ is said to be of *contact type* if there exists a vector field L, defined is some neighbourhood of Σ, which is transverse to Σ and satisfies $\mathcal{L}_L\omega = \omega$. Such a vector field is called a *Liouville* vector field.

For more details on hypersurfaces of contact type see [107].

Example 5.5.2. The sphere $S^{2n-1} \subset (\mathbb{R}^{2n}, \omega_0)$ is a hypersurface of contact type, the Liouville vector field being given by the radial vector field

$$L = \frac{1}{2}\sum_j \left(p_j \frac{\partial}{\partial p_j} + q_j \frac{\partial}{\partial q_j}\right).$$

Let Σ be a compact hypersurface of contact type with a Liouville vector field L. Define $\lambda = i_L\omega$. Then λ is a primitive of ω:

$$\mathrm{d}\lambda = \mathrm{d}i_L\omega = \mathcal{L}_L\omega = \omega,$$

where in the second equality we used Cartan's identity and the fact that $\mathrm{d}\omega = 0$. We denote by $\alpha = \lambda|_\Sigma$ its restriction to Σ.

Exercise 5.5.3. Choose $\varepsilon > 0$ so that the flow ϕ_t of L is well defined at points of Σ for all $|t| < \varepsilon$. Put

$$\Phi\colon \Sigma \times (-\varepsilon, \varepsilon) \to M, \quad (x, t) \mapsto \phi_t(x).$$

Prove that $\Phi^*\omega = \mathrm{d}(\mathrm{e}^t\alpha)$. *Hint*: Prove that $\mathcal{L}_L\lambda = \lambda$, and hence $\phi_t^*\lambda = \mathrm{e}^t\lambda$. Deduce that $\Phi^*\lambda = \mathrm{e}^t\alpha$.

Let (M, ω) be a closed symplectic manifold. Let $H \in C^\infty(M)$ be an autonomous Hamiltonian. Denote its Hamiltonian flow by h_t. Let $\Sigma = \{H = c\}$ be a regular level set of H on which the Hamiltonian flow generates a free S^1-action. Denote by T the minimal period: $h_T|_\Sigma = \mathbb{1}$. In this case, we described in Section 1.7 the construction of the reduced symplectic form σ on the quotient $N = \Sigma/S^1$. Now suppose that M carries a symplectic quasi-state $\zeta \colon C(M) \to \mathbb{R}$. Assume further that Σ is of contact type and that $\tau(\Sigma) = 1$ for the quasi-measure τ associated to ζ.

Theorem 5.5.4. *Under these conditions the quasi-state ζ induced a symplectic quasi-state $\bar\zeta$ on the reduced manifold (N, σ). Moreover, if ζ has the vanishing or* Ham-*invariance properties, so does $\bar\zeta$. If ζ satisfies the Poisson bracket inequality, so does $\bar\zeta$, with the same constant.*

Let us remark that Borman's construction works in more general settings. In particular, the contact-type assumption is unnecessary. Moreover, when ζ is associated to a subadditive spectral invariant with bounded spectral pseudo-norm, this subadditive spectral invariant gives rise to a Calabi quasi-morphism on $\mathrm{Ham}(M, \omega)$, and a similar construction yields an induced Calabi quasi-morphism on $\mathrm{Ham}(N, \sigma)$. See [26, 27] for more details.

The remainder of this section is dedicated to the proof of Theorem 5.5.4. By Exercise 5.5.3, a neighbourhood U of Σ is symplectomorphic to $\Sigma \times (-\varepsilon, \varepsilon)$, equipped with the form $\mathrm{d}(\mathrm{e}^t\alpha)$. Denote by $\pi \colon U \to \Sigma$ and $\rho \colon \Sigma \to N$ the projections, and by $\tau = \rho \circ \pi$ their composition. Choose $0 < \delta < \varepsilon$ and let $\eta \colon \mathbb{R} \to [0, 1]$ be a cutoff function which equals 1 near 0 and is supported in $(-\delta, \delta)$. Define $\theta \colon U \to \mathbb{R}$ by $\theta(x, t) = \eta(t)$. For $F \in C^\infty(N)$, we define

$$\widetilde{F} = \theta \cdot \tau^* F \colon U \to \mathbb{R},$$

and note that $\widetilde{F}$ is compactly supported inside U, and hence extends to M. Observe that the assignment $F \mapsto \widetilde{F}$ is linear and Lipschitz in the uniform norms. With this preparation we can define the reduced quasi-state

$$\bar\zeta \colon C(N) \to \mathbb{R}, \quad \bar\zeta(F) = \zeta(\widetilde{F}).$$

We split Theorem 5.5.4 into the following two propositions:

Proposition 5.5.5. *The functional $\bar\zeta$ is a symplectic quasi-state on N. Moreover, it does not depend on the choice of cutoff function θ.*

Proposition 5.5.6. *Suppose ζ has the vanishing or* Ham-*invariance properties or satisfies the Poisson bracket inequality. Then so does $\bar\zeta$.*

Exercise 5.5.7. Prove that for any $F, G \in C^\infty(N)$ the following hold:

(1) The functions $\tau^* F$ and t commute, $\{\tau^* F, t\} \equiv 0$.

(2) The Poisson bracket of $\widetilde{F}$ and $\widetilde{G}$ is given by

$$\{\widetilde{F}, \widetilde{G}\}(x, t) = \theta^2 \mathrm{e}^{-t} \{F, G\}(\rho x), \quad x \in \Sigma,\ t \in (-\varepsilon, \varepsilon). \tag{5.10}$$

(3) The Hamiltonian vector fields of F and τ^*F are related by

$$\tau_* \operatorname{sgrad} \tau^* F(x,t) = \mathrm{e}^{-t} \operatorname{sgrad} F(\rho x), \quad x \in \Sigma,\ t \in (-\varepsilon, \varepsilon). \tag{5.11}$$

PROOF OF PROPOSITION 5.5.5. Let us verify that $\bar{\zeta}$ satisfies the axioms of a symplectic quasi-state:

First we check normalization. Denote by 1 the constant function 1 on N, and note that $\widetilde{1} = \theta$. Since $\theta|_\Sigma = 1$, Proposition 5.4.22 implies that $\bar{\zeta}(1) = \zeta(\theta) = 1$.

Next, we check positivity. Let $F \in C(N)$ such that $F \geq 0$. Then clearly $\widetilde{F} \geq 0$, and hence the positivity of ζ gives $\bar{\zeta}(F) = \zeta(\widetilde{F}) \geq 0$.

Let us check quasi-linearity. Let $F, G \in C(N)$ be Poisson commuting functions. By definition, there are smooth functions $F_k, G_k \in C^\infty(N)$ such that

$$F_k \to F, \quad G_k \to G, \quad \text{and} \quad \{F_k, G_k\} \to 0,$$

in the C^0-topology. Then clearly $\widetilde{F_k} \to \widetilde{F}$ and $\widetilde{G_k} \to \widetilde{G}$ in the C^0-topology, and by (5.10) we also have $\{\widetilde{F_k}, \widetilde{G_k}\} \to 0$. In other words, the function $\widetilde{F}, \widetilde{G} \in C(M)$ Poisson commute, and hence the quasi-linearity of ζ yields

$$\bar{\zeta}(aF + bG) = \zeta(a\widetilde{F} + b\widetilde{G}) = a\zeta(\widetilde{F}) + b\zeta(\widetilde{G}) = a\bar{\zeta}(F) + b\bar{\zeta}(G).$$

Finally, the fact that $\bar{\zeta}$ does not depend on the choice of cutoff function θ follows from Proposition 5.5.8. □

Proposition 5.5.8. *Let $F, G \in C^\infty(M)$ be smooth functions supported in U which Poisson commute with t and such that $F|_\Sigma = G|_\Sigma$. Then $\zeta(F) = \zeta(G)$.*

PROOF. Let $\psi\colon \mathbb{R} \to [0,1]$ be a cutoff function supported in $(-\varepsilon, \varepsilon)$ with $\psi(0) = 1$, and define $\chi\colon U \to \mathbb{R}$ by $\chi(x,t) = \psi(t)$. Then since F commutes with t, $\{F, \chi F\} \equiv 0$, and hence the quasi-linearity of ζ and Proposition 5.4.22 give

$$\zeta(F) - \zeta(\chi F) = \zeta((1-\chi)F) = 0,$$

since $(1-\chi)F$ vanishes on Σ. Therefore, $\zeta(\chi F) = \zeta(F)$ and similarly $\zeta(\chi G) = \zeta(G)$. Since ζ is Lipschitz we get

$$|\zeta(F) - \zeta(G)| = |\zeta(\chi F) - \zeta(\chi G)| \leq \|\chi(F-G)\|. \tag{5.12}$$

Now let ψ_k be a sequence of such cutoff functions, with ψ_k supported in $(-1/k, 1/k)$, and let χ_k be the corresponding functions defined on U. Since $F - G$ vanishes on Σ, we get

$$\|\chi_k(F-G)\| \xrightarrow[k\to\infty]{} 0,$$

hence by (5.12), $\zeta(F) = \zeta(G)$. □

PROOF OF PROPOSITION 5.5.6. First we introduce the following notation: if $h_t\colon N \to N$ is the Hamiltonian flow of a Hamiltonian H_t on N, we denote by $\tilde{h}_t$ the Hamiltonian flow of $\widetilde{H}_t$ on M. Observe that from (5.11) it follows that

$$\tau_* \operatorname{sgrad} \widetilde{H}_t(x,0) = \operatorname{sgrad} H_t(x), \quad x \in \Sigma,$$

and hence

$$(h_t \circ \tau)|_\Sigma = (\tau \circ \tilde{h}_t)|_\Sigma. \tag{5.13}$$

Assume that ζ has the vanishing property. Let $F \in C^\infty(N)$ have displaceable support, and let $h \in \operatorname{Ham}(N)$ be a displacing Hamiltonian diffeomorphism, and assume that $h = h_1$ is the time-1 map of the Hamiltonian flow of $H_t \colon N \to \mathbb{R}$. Then (5.13) implies that $\tilde{h}_1$ displaces $\rho^{-1}(\operatorname{supp} F) \times \{0\} \subset M$. Note that $\operatorname{supp} \widetilde{F} \subset \rho^{-1}(\operatorname{supp} F) \times (-\delta, \delta)$, where δ is the parameter from the definition of θ. Since $\bar{\zeta}$ does not depend on the choice of the cutoff function θ, we may choose δ small enough so that $\tilde{h}_1$ displaces $\operatorname{supp} \widetilde{F}$. Therefore, by the vanishing property of ζ, we get $\bar{\zeta}(F) = \zeta(\widetilde{F}) = 0$. Thus $\bar{\zeta}$ has the vanishing property.

Assume next that ζ is Ham-invariant. Let $F \in C^\infty(N)$ and let $h \in \operatorname{Ham}(N)$ be a Hamiltonian diffeomorphism generated by $H_t \colon N \to \mathbb{R}$. Observe that, by (5.13), on Σ we have

$$\widetilde{F \circ h} = \theta \cdot F \circ h \circ \tau = \theta \cdot F \circ \tau \circ \tilde{h} = \widetilde{F} \circ \tilde{h}.$$

Therefore, using Proposition 5.5.8 and the Ham-invariance of ζ, we get

$$\bar{\zeta}(F \circ h) = \zeta(\widetilde{F \circ h}) = \zeta(\widetilde{F} \circ \tilde{h}) = \zeta(\widetilde{F}) = \bar{\zeta}(F),$$

proving that $\bar{\zeta}$ is Ham-invariant.

Finally, suppose that ζ satisfies the Poisson bracket inequality with constant C. Let $F, G \in C^\infty(N)$. Using (5.10) and the Poisson bracket inequality for ζ we get

$$\begin{aligned} |\bar{\zeta}(F+G) - \bar{\zeta}(F) - \bar{\zeta}(G)| &= |\zeta(\widetilde{F} + \widetilde{G}) - \zeta(\widetilde{F}) - \zeta(\widetilde{G})| \le C\sqrt{\|\{\widetilde{F}, \widetilde{G}\}\|} \\ &= C\sqrt{\theta^2 \mathrm{e}^{-t}\|\{F,G\}\|} \le \mathrm{e}^{\delta/2} C\sqrt{\|\{F,G\}\|}. \end{aligned}$$

Thus $\bar{\zeta}$ satisfies the Poisson bracket inequality, with constant $\mathrm{e}^{\delta/2}C$. Since $\bar{\zeta}$ does not depend on the choice of cutoff function θ, we may take $\delta \to 0$, and conclude that the inequality is satisfied with constant C. □

Example 5.5.9. Recall from Section 4.7 that $\mathbb{C}P^2$ admits a symplectic quasi-state ζ associated to a subadditive spectral invariant with bounded spectral pseudo-norm. In particular, ζ satisfies the Poisson bracket inequality and has the vanishing and Ham-invariance properties.

Exercise 5.5.10. This exercise is borrowed from [107, Example 7.14]. Identify

$$\mathbb{C}P^1 = \{[0 : z_1 : z_2]\} \subset \mathbb{C}P^2.$$

Let B^4 denote the open unit ball in $(\mathbb{R}^4, \omega_0)$. Prove that the map

$$\Phi \colon \mathbb{C}P^2 \setminus \mathbb{C}P^1 \to B^4, \quad [z_0 : z_1 : z_2] \mapsto \left(\frac{z_1}{\sqrt{z_0^2 + z_1^2 + z_2^2}}, \frac{z_2}{\sqrt{z_0^2 + z_1^2 + z_2^2}} \right)$$

is a symplectomorphism.

Now let $L \subset \mathbb{C}P^2$ be the *Clifford torus:*

$$L = \{|z_0| = |z_1| = |z_2|\} \subset \mathbb{C}P^2.$$

Clearly, $L \subset \mathbb{C}P^2 \setminus \mathbb{C}P^1$ and it is easily verified that $\Phi(L) = S^1(1/\sqrt{3}) \times S^1(1/\sqrt{3}) \subset B^4$, where $S^k(r)$ denotes the k-sphere of radius r. In particular, $\Phi(L) \subset S^3(\sqrt{2/3})$. Denote $S = \Phi^{-1}(S^3(\sqrt{2/3})) \subset \mathbb{C}P^2$. It follows from Example 5.5.2 that $S \subset \mathbb{C}P^2$ is a hypersurface of contact type, and as we saw in Example 1.7.3 it admits a Hamiltonian S^1 action whose Marsden–Weinstein reduction yields the projective space $\mathbb{C}P^1$, equipped with the Fubini–Study form. Moreover, as we shall show in the following chapter (see the discussion after Exercise 6.1.17), $\tau(L) = 1$ for the quasi-measure τ associated to ζ, and since $S \supset L$, $\tau(S) = 1$ as well. It follows that ζ induces a symplectic quasi-state $\bar{\zeta}$ on the reduced manifold $\mathbb{C}P^1 \simeq S^2$. Furthermore, $\bar{\zeta}$ satisfies the Poisson bracket inequality and has the vanishing and Ham-invariance properties. We conclude, using Exercise 5.4.29, that the reduced quasi-state $\bar{\zeta}$ is the median quasi-state.

5.6. Lie quasi-states

Let $\mathfrak{g}$ be a Lie algebra over $\mathbb{R}$. A function $\zeta\colon \mathfrak{g} \to \mathbb{R}$ is called *a Lie quasi-state* [62] if it is linear on every abelian subalgebra. Interestingly enough, Lie quasi-states with mild extra analytic properties exhibit quite rigid behavior. For instance, Gleason's Theorem 5.2.4 readily yields the following result:

Theorem 5.6.1. *Every Lie quasi-state on the unitary algebra $\mathfrak{u}(n)$ with $n \geq 3$ which is bounded in a neighborhood of 0 is necessarily linear.*

PROOF. Identify $\mathfrak{u}(n)$ with the space of all Hermitian operators on $\mathbb{C}^n$. Recall that instead of the boundedness of ζ near zero Gleason assumes that $\zeta(\mathbb{1}) = 1$ and ζ is nonnegative on the set of positive operators. To obtain this nonnegativity condition from the boundedness near zero, one just needs to modify ζ to

$$\zeta'(A) := (\zeta(\mathbb{1}) + Cn)^{-1} \cdot (\zeta(A) + C \operatorname{tr} A)$$

for a sufficiently large $C > 0$. The nonnegativity of ζ' follows from the fact that for a positive Hermitian operator $\operatorname{tr} A \geq \|A\|_{\text{op}}$, where $\|A\|_{\text{op}}$ is the operator norm. Applying Gleason's Theorem we get that ζ', and therefore also ζ, is linear. □

What happens on other Lie algebras? A step in this direction was made in [62] for the symplectic algebra $\mathfrak{g} = \mathfrak{sp}(2n, \mathbb{R})$. The space of Lie quasi-states which are bounded near 0 is in this case infinite-dimensional. However, rigidity is regained when one restricts oneself to *continuous* Lie quasi-states. Denote by $\mathcal{Q}(\mathfrak{g})$ the quotient of the space of all continuous Lie quasi-states on $\mathfrak{g}$ by the Lie coalgebra $\mathfrak{g}^*$.

Theorem 5.6.2 ([62]). *For $n \geq 3$, $\dim \mathcal{Q}(\mathfrak{sp}(2n, \mathbb{R})) = 1$.*

The proof given in [62] involves the (easier) continuous version of Theorem 5.6.1 combined with a lengthy and technical argument. Finding a conceptual proof is still a challenge. For the sake of completeness, let us mention that the statement of Theorem 5.6.2 becomes wrong for $n = 1$, and the problem is still open for $n = 2$. Lie quasi-states on other Lie algebras remain basically unexplored.

On the more optimistic note, the generator of $\mathcal{Q}(\mathfrak{sp}(2n, \mathbb{R}))$ admits a simple and conceptual description. We shall present it in the following more general context. Let μ be a homogeneous quasi-morphism on a Lie group G. Then the function ζ on the Lie algebra $\mathfrak{g}$ defined by $\zeta(x) = \mu(\exp(x))$ is a Lie quasi-state. Indeed, if $[x, y] = 0$, then $\exp(x)$ and $\exp(y)$ commute and $\exp(x + y) = \exp(x)\exp(y)$. Since μ is a homomorphism when restricted to an abelian subgroup (see Proposition 3.1.4 above), we get that ζ is linear on the span of x and y, as required. In particular, the generator of $\mathcal{Q}(\mathfrak{sp}(2n, \mathbb{R}))$ comes from the Maslov quasi-morphism on $\widetilde{\mathrm{Sp}}(2n)$ (see Definition 3.4.14 above).

Recall now that the space $\mathcal{H}$ of smooth functions with zero mean on a symplectic manifold can be identified with the Lie algebra of the group $\widetilde{\mathrm{Ham}}(M, \omega)$. Given a subadditive spectral invariant with bounded spectral pseudo-norm, we get a symplectic quasi-state ζ on $\mathcal{H}$ and a quasi-morphism σ on $\widetilde{\mathrm{Ham}}$. Clearly, ζ is a Lie quasi-state on $\mathcal{H}$ which is obtained from σ as its pull-back under the exponential map.

CHAPTER 6

Applications of Partial Symplectic Quasi-states

In this chapter we present applications of partial symplectic quasi-states to symplectic topology. These include symplectic intersection theory, Lagrangian knots, and Hofer's geometry.

6.1. Symplectic intersections

In this section we discuss some classes of nondisplaceable subsets detected by the function theory on symplectic manifolds. Our exposition follows [59] and [63].

6.1.1. The nondisplaceable fiber theorem. Let $\mathcal{A} \subset C^\infty(M)$ be a finite-dimensional Poisson commutative subspace: $\{F, G\} = 0$ for all $F, G \in \mathcal{A}$. Such subspaces play an important role in Hamiltonian dynamics, see Section 6.1.3 below. Interestingly enough, they often serve as a source of interesting examples of nondisplaceable subsets.

Define the *moment map* $\Phi\colon M \to \mathcal{A}^*$, where $\mathcal{A}^*$ is the dual space of $\mathcal{A}$, by

$$\langle \Phi(x), H \rangle = H(x) \quad \forall H \in \mathcal{A}.$$

Choose a basis $F_1, \ldots, F_N$ of $\mathcal{A}$. Then in coordinates induced by the dual basis, Φ has the form

$$\Phi\colon x \mapsto \big(F_1(x), \ldots, F_N(x)\big) \in \mathbb{R}^N.$$

Denote by $\Delta \subset \mathcal{A}^*$ the image of the moment map Φ. For $p \in \Delta$, the set $\Phi^{-1}(p)$ is called the *fiber* of Φ over p.

Definition 6.1.1. A fiber $\Phi^{-1}(p)$ is called a *stem* if all other fibers of Φ are displaceable.

Example 6.1.2. On $S^2 \subset \mathbb{R}^3$ consider a height function $F(x, y, z) = z$, and let $\mathcal{A}$ be the subspace of $C^\infty(M)$ spanned by F. The fibers of the moment map are simply level sets of F, and the equator is a stem.

We say that $X \subset M$ is a stem if it is a stem of some Poisson commutative subspace. Note that the image of a stem under a symplectomorphism is again a stem.

Example 6.1.3. Let $X \subset M$ be a closed subset whose complement $M \setminus X$ has a finite number of connected components U_i with the following property: *Every compact subset of U_i is displaceable.* (For instance, this is the case when X is the codimension 1 skeleton of a sufficiently fine triangulation of M.) Then X is a stem. Indeed, for each connected component U_i, $i = 1, \ldots, N$, of $M \setminus X$ choose a function $F_i \colon U_i \to [0,1]$ which is strictly positive on the interior of U_i and vanishes outside U_i. Observe that F_i attains its minimum on $M \setminus U_i$ and hence its differential on this set vanishes. Since for $i \neq j$

$$(M \setminus U_i) \cup (M \setminus U_j) = M,$$

the functions F_i and F_j Poisson commute. Further, given $p = (p_1, \ldots, p_N) \in [0,1]^N$ with $p_i > 0$ for some i, the fiber $\Phi^{-1}(p)$ lies in U_i and hence is displaceable. Thus $X = \Phi^{-1}(0)$ is a stem.

Exercise 6.1.4. Prove that a product of two stems is a stem. *Hint*: Let $X_i \subset M_i$ be a stem associated to a finite-dimensional Poisson commutative subspace $\mathcal{A}_i \subset C^\infty(M_i)$, $i = 1, 2$. Denote by $\theta_i \colon M_1 \times M_2 \to M_i$ the natural projection. Prove that $X_1 \times X_2$ is a stem of the Poisson commutative subspace $\mathrm{Span}_{\mathbb{R}}(\theta_1^* \mathcal{A}_1 \cup \theta_2^* \mathcal{A}_2)$ of $C^\infty(M_1 \times M_2)$.

Stems play an important role in our story because of the following result.

Theorem 6.1.5. *Any two stems intersect.*

The proof will be given in the next subsection. In particular, since all the fibers of the moment map are disjoint, if a finite-dimensional Poisson commutative subspace possesses a stem, it is necessarily unique (let us point out, however, that a stem does not always exist, see Remark 6.2.7 for an example).

Corollary 6.1.6. *Stems are nondisplaceable by* $\mathrm{Symp}(M, \omega)$.

PROOF. Let $X \subset M$ be a stem. Then for any $\varphi \in \mathrm{Symp}(M, \omega)$, $\varphi(X)$ is again a stem and hence $\varphi(X) \cap X \neq \varnothing$. □

Note that nondisplaceability by symplectomorphisms is stronger than the standard notion of nondisplaceability by Hamiltonian diffeomorphisms.

Example 6.1.7. It follows from Example 6.1.2 and Exercise 6.1.4 that the *Clifford torus*, which is the product of equators in $S^2 \times \cdots \times S^2$ equipped with the split symplectic form, is a stem, and hence nondisplaceable by symplectomorphisms. Nondisplaceability of the Clifford torus by Hamiltonian diffeomorphisms was proved earlier by Oh [117] with the use of Lagrangian Floer theory. Replacing equators by 1-skeletons of sufficiently fine triangulations of S^2 as in Example 6.1.3, we get nondisplaceability of a *singular* Lagrangian submanifold, a phenomenon which at the moment is not covered by Lagrangian Floer theory (see Figure 6.1).

Theorem 6.1.8 (Nondisplaceable fiber theorem). *Every finite-dimensional Poisson commutative subspace* $\mathcal{A} \subset C^\infty$ *has a nondisplaceable fiber.*

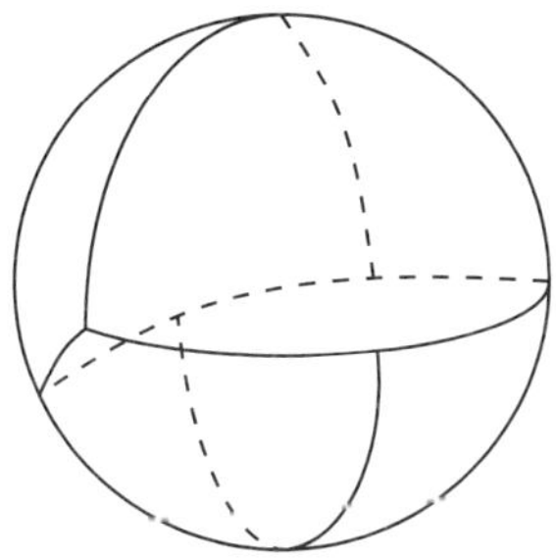

FIGURE 6.1. A singular Lagrangian in S^2

PROOF. Assume, on the contrary, that all the fibers of $\mathcal{A}$ are displaceable. Then there are at least two distinct (and therefore disjoint) fibers. Since both are stems, they must intersect. We get a contradiction. □

Let us present an immediate consequence of the nondisplaceable fiber theorem which will play an important role in Chapter 9 below.

Corollary 6.1.9. *No partition of unity subordinated to a finite cover of a closed symplectic manifold by displaceable open sets can consist of Poisson commuting functions.*

PROOF. Assume, on the contrary, that the functions entering such a partition of unity span a Poisson commutative subspace. Every fiber of its moment map is contained in at least one set of the cover, and hence it is displaceable. We got a contradiction with the nondisplaceable fiber theorem. □

6.1.2. Superheavy subsets. Let (M, ω) be a closed symplectic manifold. Fix a subadditive spectral invariant c on $\widetilde{\mathrm{Ham}}(M, \omega)$. Let $\zeta \colon C(M) \to \mathbb{R}$ be the corresponding partial symplectic quasi-state (see Section 4.5 above).

Definition 6.1.10 ([63]). A closed subset $X \subset M$ is called *superheavy* if $\zeta(F) = 0$ for any smooth function F vanishing on X.

Observe that since ζ is continuous in the uniform norm, X is superheavy provided $\zeta(F) = 0$ for every F which vanishes *near* X. Further, since $\zeta(F + a) = \zeta(F) + a$, we have that $\zeta(F) = a$ for every F with $F|_X = a$.

Example 6.1.11. Assume that the spectral pseudo-norm associated to c is bounded, and thus ζ is a genuine symplectic quasi-state. Then a closed subset $X \subset M$ is superheavy if and only if $\tau(X) = 1$, where τ is the quasi-measure associated to ζ. This follows from Proposition 5.4.22 above.

Proposition 6.1.12. *Every two superheavy sets intersect.*

PROOF. Assume, on the contrary, that X and Y are two disjoint superheavy subsets. Choose a smooth function F on M which equals 0 on X and equals 1 on Y. By superheaviness, $0 = \zeta(F) = 1$, a contradiction. □

Theorem 6.1.5 stating that every two stems intersect is an immediate consequence of Proposition 6.1.12 and the next result:

Proposition 6.1.13. *Every stem is superheavy.*

PROOF. Let X be a stem of a finite-dimensional Poisson commutative subspace generated by the functions $F_1, \ldots, F_N$. Put $\Phi = (F_1, \ldots, F_N)\colon M \to \mathbb{R}^N$. Assume that $X = \Phi^{-1}(p)$ for some $p \in \Delta = \Phi(M)$.

(1) Let $H\colon \mathbb{R}^N \to \mathbb{R}$ be any function which vanishes in a neighborhood, say V, of p. We claim that $\zeta(\Phi^* H) = 0$. Indeed, consider a cover of $\Delta \setminus V$ by sufficiently small open subsets $U_1, \ldots, U_d$ so that $\Phi^{-1}(U_i)$ is displaceable for all i. Such a cover exists since $\Delta \setminus V$ is compact and all fibers of Φ except maybe p are displaceable. Choose functions $\rho_1, \ldots, \rho_d\colon \mathbb{R}^N \to \mathbb{R}$ such that $\operatorname{supp}(\rho_i) \subset U_i$ and $\sum_i \rho_i = 1$ on $\Delta \setminus V$. Write $H = \sum_i \rho_i H$.

Exercise 6.1.14. Show that for any $A, B \in C^\infty(\mathbb{R}^N)$, $\{A \circ \Phi, B \circ \Phi\} \equiv 0$. *Hint*: Use that F_i pairwise commute.

By the exercise, the pullbacks $K_i := \Phi^*(\rho_i H)$ pairwise commute. The supports of K_i are displaceable and hence have finite spectral width. We have that $\zeta(K_i) = 0$ for all i by the vanishing property of ζ. Furthermore, for all $m = 1, \ldots, d-1$,

$$\zeta(K_1 + \cdots + K_m + K_{m+1}) = \zeta(K_1 + \cdots + K_m)$$

by partial quasi-additivity. Therefore

$$\zeta(\Phi^* H) = \zeta\left(\sum_{i=1}^{d} K_i\right) = 0,$$

which proves the claim.

(2) It remains to notice that if G is any function on M which vanishes near X, then there exist functions $H_-, H_+\colon \mathbb{R}^N \to \mathbb{R}$ which vanish near p such that

$$\Phi^* H_- \leq G \leq \Phi^* H_+.$$

By monotonicity of ζ and step 1 of the proof,

$$0 = \zeta(\Phi^* H_-) \leq \zeta(G) \leq \zeta(\Phi^* H_+) = 0,$$

and hence $\zeta(G) = 0$. Thus X is superheavy. □

6.1.3. Examples of Poisson commutative subspaces.

Example 6.1.15 (Hamiltonian torus actions). Suppose $\mathbb{T}^k$ acts on (M, ω) by Hamiltonian diffeomorphisms. This induces a homomorphism of Lie algebras

$$\operatorname{Lie}(\mathbb{T}^k) \to \operatorname{Lie}\big(\operatorname{Ham}(M, \omega)\big) = \mathcal{H}$$

whose image is a Poisson commutative subspace. See [107, Chapter 5] for more details on Hamiltonian actions by general Lie groups.

Exercise 6.1.16. Consider the action of $\mathbb{T}^2 = S^1 \times S^1$ on $S^2 \times S^2$ by rotation about the vertical axis in each factor. Check that the image of the moment map is the square $[-1,1] \times [-1,1] \subset \mathbb{R}^2$ and that the inverse image of the point $(0,0)$ is the Clifford torus, mentioned in Example 6.1.7.

Exercise 6.1.17. Define an action of $\mathbb{T}^n$ on $\mathbb{C}P^n$ by

$$(t_1, \dots, t_n) \cdot [z_0 : z_1 : \dots : z_n] = [z_0 : \mathrm{e}^{2\pi i t_1} z_1 : \dots : \mathrm{e}^{2\pi i t_n} z_n].$$

Show that the image of the moment map is the standard n-simplex. Show that the preimage of the barycentre is

$$L = \{|z_0| = |z_1| = \dots = |z_n|\}.$$

Prove that L is a stem. *Hint*: Look at the action of the symmetric group S_{n+1} on $\mathbb{C}P^n$ by permutations of the homogeneous coordinates.

The torus L is called the *Clifford torus* in $\mathbb{C}P^n$. As we shall see in Example 12.6.3 below, $\mathbb{C}P^n$ admits a spectral invariant with bounded spectral pseudo-norm. Denote by τ the corresponding quasi-measure. It follows from Example 6.1.11 and Proposition 6.1.13 above that $\tau(L) = 1$.

The moment maps of Hamiltonian actions of the torus $\mathbb{T}^n$ on closed symplectic manifolds of dimension $2n$ have a very special structure. In order to describe it, we need the following important definition. Recall that a subspace L of a symplectic vector space (E, ω) is called *Lagrangian* if $\dim L = \frac{1}{2} \dim E$ and ω vanishes on L.

Definition 6.1.18. Let (M, ω) be a symplectic manifold. A submanifold $L \subset M$ is called *Lagrangian* if for any $x \in L$, $T_x L \subset T_x M$ is a Lagrangian subspace.

Theorem 6.1.19 (Atiyah [11], Guillemin–Sternberg [82]). *Let (M^{2n}, ω) be a closed connected symplectic manifold of dimension $2n$, and suppose that it admits a Hamiltonian action of $\mathbb{T}^n$ with moment map $\Phi\colon M \to \mathbb{R}^n$. Then the image of Φ is a convex polytope, and the inverse image of every point in its interior is a Lagrangian torus.*

Hamiltonian actions of the torus $\mathbb{T}^n$ on symplectic manifolds M^{2n} form a particular case of the following interesting class of Hamiltonian systems.

Example 6.1.20 (Integrable systems). Poisson commutative subspaces naturally appear in the context of *integrable systems* of classical mechanics, see [7, Chapter 10]. Let (M^{2n}, ω) be a closed symplectic manifold. A Hamiltonian $H_1\colon M \to \mathbb{R}$ is called integrable if there exists functions $H_2, \dots H_n \in C^\infty(M)$ such that $\{H_i, H_j\} = 0$ for all $i, j = 1, \dots, n$ and moreover the regular levels of the corresponding moment map

$$\Phi\colon M \to \mathbb{R}^n, \quad x \mapsto \big(H_1(x), \dots, H_n(x)\big),$$

are dense in M (our definition is weaker than the one in [7]). The importance of this class of systems is due to the fact that the regular levels of Φ are invariant under the Hamiltonian flow of H_1 and carry a simple dynamics.

Let us explain this in more detail. We denote the Hamiltonian flows of the H_i by h_i^t. By Exercise 1.3.11, these flows commute, that is, $h_i^t h_j^s = h_j^s h_i^t$ for any $s, t \in \mathbb{R}$. This allows us to define an $\mathbb{R}^n$-action on M: for $\mathrm{v} = (\mathrm{v}_1, \ldots, \mathrm{v}_n) \in \mathbb{R}^n$ we set

$$h^{\mathrm{v}} x = h_1^{\mathrm{v}_1} \cdots h_n^{\mathrm{v}_n} x.$$

Now, let $p \in \mathbb{R}^n$ be a regular value of Φ with $\Phi^{-1}(p) \neq \varnothing$. Let L be a connected component of $\Phi^{-1}(p)$.

Proposition 6.1.21. *L is a Lagrangian torus invariant under the $\mathbb{R}^n$-action.*

The proof we give is taken from [7].

PROOF. Pick any point $x \in L$. Denote by Σ_i the level set of H_i passing through x. Since p is a regular value of Φ, we have that $\dim L = n$ and

$$T_x L = \bigcap_{i=1}^{n} T_x \Sigma_i.$$

Furthermore, $0 = \{H_i, H_j\} = dH_i(\operatorname{sgrad} H_j)$, and therefore the field $\operatorname{sgrad} H_j$ is tangent to Σ_i for all i and j. It follows that

$$T_x L = \operatorname{span}\{\operatorname{sgrad} H_1(x), \ldots, \operatorname{sgrad} H_n(x)\}.$$

This implies that L is h_i-invariant for any $1 \leq i \leq n$, and hence invariant under the $\mathbb{R}^n$-action. Additionally, since $0 = \{H_i, H_j\} = \omega(\operatorname{sgrad} H_j, \operatorname{sgrad} H_i)$, $\omega|_L = 0$ and hence L is Lagrangian. Finally, let us show that $L \simeq \mathbb{T}^n$. Fix a point $x_0 \in L$. Consider its stabilizer,

$$\Gamma := \operatorname{stab} x_0 = \{\mathrm{v} \in \mathbb{R}^n : h^{\mathrm{v}} x_0 = x_0\}.$$

The following three claims are left as exercises:

(1) The action on L is transitive.
(2) Γ does not depend on $x_0 \in L$.
(3) $\Gamma \subset \mathbb{R}^n$ is discrete.

It follows that

$$\Gamma = \operatorname{span}_{\mathbb{Z}}\{\xi_1, \ldots, \xi_k\}$$

for independent (over $\mathbb{R}$) vectors $\xi_1, \ldots, \xi_k \in \mathbb{R}^n$, and that $L \simeq \mathbb{R}^n/\Gamma$. Since L is compact, this implies that $k = n$ and hence $L \simeq \mathbb{R}^n/\mathbb{Z}^n = \mathbb{T}^n$. □

6.2. Lagrangian knots

Let (M, ω) be a closed symplectic manifold and let $L \subset M$ be a Lagrangian submanifold. We associate to L two classical invariants, which are homomorphisms on the relative homotopy group $\pi_2(M, L)$:

The *symplectic area class* is defined by

$$[\omega] \colon \pi_2(M, L) \to \mathbb{R}, \quad [D] \mapsto \int_D \omega.$$

The *Maslov class*

$$\mu\colon \pi_2(M,L) \to \mathbb{Z},$$

is defined as follows. Let $[u] \in \pi_2(M,L)$, represented by $u\colon D^2 \to M$ with $u(S^1) \subset L$. The bundle u^*TM has a contractible base, and hence can be trivialized as a symplectic vector bundle. Thus u^*TM is isomorphic to the trivial bundle $D^2 \times (\mathbb{R}^{2n}, \omega_0)$. By using the trivialization one identifies, for each $x \in S^1$, the subspace $T_{u(x)}L$ of the fiber $(u^*TM)_x$ with a Lagrangian subspace of $(\mathbb{R}^{2n}, \omega_0)$. Thus we obtain a map $\tilde{u}\colon S^1 \to \Lambda(\mathbb{R}^{2n})$, where $\Lambda(\mathbb{R}^{2n})$ is the Lagrangian Grassmannian introduced in Section 3.4. The Maslov class of $[u]$ is defined to be the Maslov index of the loop $\tilde{u}$ (see Definition 3.4.11):

$$\mu([u]) = \mathrm{Maslov}(\tilde{u}).$$

This definition does not depend on the trivialization, since any other trivialization gives a loop in $\Lambda(\mathbb{R}^{2n})$ which differs from $\tilde{u}$ by a contractible loop of symplectic matrices. For more details see [107, Section 2.3].

Definition 6.2.1. A Lagrangian submanifold $L \subset M$ is called *(spherically) monotone* if there exists $\kappa > 0$ such that $[\omega] = \kappa\mu$.

Exercise 6.2.2. Take the two-sphere S^2 equipped with an area form ω. Let $L \subset S^2$ be the equator. Prove that $L \subset (S^2, \omega)$ and $L \times L \subset (S^2 \times S^2, \omega \oplus \omega)$ are monotone Lagrangian submanifolds.

Given two diffeomorphic monotone Lagrangian submanifolds L_0, L_1 of M with the same monotonicity constant κ, are they Hamiltonian isotopic, that is, does there exist $\varphi \in \mathrm{Ham}(M)$ such that $\varphi(L_0) = L_1$? This is a particular case of the *Lagrangian knots* problem which goes back to Arnold's seminal survey [8].

Example 6.2.3. A *local* Lagrangian knot in $(\mathbb{R}^{2n}, \mathrm{d}p \wedge \mathrm{d}q)$ is an embedded Lagrangian $\mathbb{R}^n$ which coincides with the linear Lagrangian subspace $L_0 = \{p = 0\}$ outside a compact set. In [54] it is proved that there are no nontrivial local Lagrangian knots in $\mathbb{R}^4$: all of them are Hamiltonian isotopic to L_0. In higher dimensions the problem of local Lagrangian knots is largely open.

The first examples of nontrivial Lagrangian knots are due to Chekanov [46]. He found monotone Lagrangian tori in $\mathbb{R}^{2n}$ with equal monotonicity constants which are not Hamiltonian isotopic.

The main message of this section is that in certain situations Lagrangian knots can be distinguished with the help of symplectic quasi-measures. We illustrate this in the next example. Proofs of the results discussed below are too involved to be presented in our book, so we give a brief survey only.

Example 6.2.4 (Coupled spins). Let M be a product of two copies of S^2:

$$M = S_1^2 \times S_2^2, \qquad S_i^2 = \{x_i^2 + y_i^2 + z_i^2\} \subset \mathbb{R}^3, \quad i = 1, 2,$$

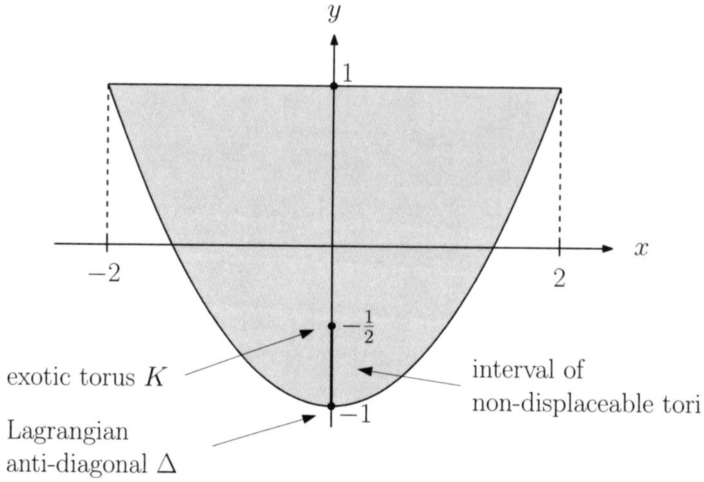

FIGURE 6.2. Image of the moment map

equipped with the split symplectic form $\omega \oplus \omega$, where ω is the standard area form on S^2. Let $H_1, H_2 \colon M \to \mathbb{R}$,

$$H_1 = z_1 + z_2, \quad H_2 = x_1x_2 + y_1y_2 + z_1z_2.$$

Exercise 6.2.5. Prove that $\{H_1, H_2\} = 0$. *Hint*: Use Exercise 1.3.13.

Note that

$$\begin{aligned} 2 + 2H_2 &= x_1^2 + y_1^2 + z_1^2 + x_2^2 + y_2^2 + z_2^2 + 2(x_1x_2 + y_1y_2 + z_1z_2) \\ &= (x_1 + x_2)^2 + (y_1 + y_2)^2 + (z_1 + z_2)^2 \geq (z_1 + z_2)^2 = H_1^2. \end{aligned}$$

It follows that the image of the moment map is the region

$$\{2 + 2y \geq x^2, \ -2 \leq x \leq 2, \ -1 \leq y \leq 1\} \subset \mathbb{R}^2$$

(see Figure 6.2).

Note that $(0, -1)$ is a critical value of the moment map. The corresponding fiber $\Delta = \Phi^{-1}(0, -1)$ is given by the equations

$$\Delta = \{x_1x_2 + y_1y_2 + z_1z_2 = -1, z_1 = -z_2\} = \{(r_1, r_2) \in S^2 \times S^2 : r_2 = -r_1\}.$$

We see that Δ is the anti-diagonal in $S^2 \times S^2$ and hence it is diffeomorphic to the two-sphere, as opposed to regular fibers, which are tori.

Consider now the following regular fiber of the moment map: $K = \Phi^{-1}(0, -\frac{1}{2})$. We leave it as an exercise to verify that K is a monotone Lagrangian torus with the same monotonicity constant as the Clifford torus L.

Theorem 6.2.6 ([55]). *There exist quasi-measures τ_+, τ_- on $S^2 \times S^2$, associated to symplectic quasi-states, such that*

(1) $\tau_-(\Delta) = 1$, $\tau_-(K) = 0$.
(2) $\tau_+(\Delta) = 0$, $\tau_+(K) = 1$.

However, as we saw, L is a stem, and hence $\tau_-(L) = 1$. Since $\tau_-(K) = 0$, we see that *Lagrangian tori K and L are not Hamiltonian isotopic, although they have identical classical invariants.* We refer to the end of Chapter 12 for further information about the quasi-measure τ_-.

Remark 6.2.7. From Theorem 6.2.6 it follows that both Δ and K are nondisplaceable, and thus the subspace spanned by H_1 and H_2 has no stems.

Finally, let us mention that for any $-1 < \lambda < -\frac{1}{2}$, the torus $\Phi^{-1}(0, \lambda)$ is nondisplaceable [69].

6.3. Applications to Hofer's geometry

Let (M, ω) be a closed symplectic manifold, and let c be a subadditive spectral invariant. Write σ for the map $\widetilde{\mathrm{Ham}}(M, \omega) \to \mathbb{R}$, $\sigma(\phi) = \lim_{k\to\infty} c(\phi^k)/k$. Let $\zeta\colon C^\infty(M) \to \mathbb{R}$ be the partial symplectic quasi-state associated to c, that is $\zeta(F) = \sigma(\phi_F) + \langle F \rangle$, where $\phi_F \in \widetilde{\mathrm{Ham}}(M, \omega)$ is the element generated by F, $\sigma(\phi_F) = \lim_{k\to\infty} c(\phi_F^k)/k$, and $\langle F \rangle$ stands for the mean value of F. By using the stability property of subadditive spectral invariants (see Proposition 4.3.4), we get that

$$\tilde{d}(\mathbb{1}, \phi) \geq |\sigma(\phi)| \quad \forall \phi \in \widetilde{\mathrm{Ham}}(M, \omega). \tag{6.1}$$

This inequality yields several applications to Hofer's geometry.

6.3.1. Growth of one-parameter subgroups. Let $\{f_t\}$, $t \in \mathbb{R}$, be a one-parameter subgroup of $\mathrm{Ham}(M, \omega)$, that is, a path generated by an autonomous normalized Hamiltonian F. Denote by $\{\tilde{f}_t\}$ its lift to $\widetilde{\mathrm{Ham}}(M, \omega)$. We are interested in the following question:

Question 1. What are the possible growth types of the function $t \mapsto \tilde{d}(\mathbb{1}, \tilde{f}_t)$, where $\tilde{d}$ is Hofer's pseudo-metric on the universal cover $\widetilde{\mathrm{Ham}}(M, \omega)$?

Since by time rescaling (Proposition 1.3.5) $\tilde{f}_t$ is generated by tF, we have

$$\tilde{d}(\mathbb{1}, \tilde{f}_t) \leq t\|F\|. \tag{6.2}$$

By the triangle inequality, $r_t(F) := \tilde{d}(\mathbb{1}, \tilde{f}_t)$ is a subadditive function of t, and hence the limit

$$r(F) := \lim_{t\to+\infty} \frac{r_t(F)}{t}$$

exists. We say that a one-parameter subgroup $\{\tilde{f}_t\}$ is *undistorted* on the universal cover of Ham with respect to the Hofer metric if $r(F) > 0$. Distortion of one-parameter subgroups has been studied on various occasions in connection to Hamiltonian dynamics and ergodic theory, see, e.g., [132, Chapters 8, 11].

At the moment only two patterns of asymptotic behavior of one-parameter subgroups is known: either $\{f_t\}$ is undistorted, or $r_t(F) \leq C$ for all t, in which case we say that *$\{f_t\}$ is bounded.*

Exercise 6.3.1. Show that every symplectic manifold admits a one-parameter subgroup $\{f_t\}$ which possesses a time-reversing symmetry (cf. Section 3.2 above): there exists $\phi \in \text{Ham}(M, \omega)$ such that $\phi f_t \phi^{-1} = f_t^{-1}$ for all $t \in \mathbb{R}$. Prove that $\{f_t\}$ is necessarily bounded.

Sikorav [149] proved that every one-parameter subgroup of $\widetilde{\text{Ham}}(\mathbb{R}^{2n}, \omega_0)$ is bounded. A dichotomy theorem of [136] states that on open surfaces of infinite area these are the only possibilities. However, no complete answer to Question 1 is available even for the simplest closed symplectic manifolds.

Let (M, ω) be a closed symplectic manifold. Recall that in this case $\mathcal{H}$ stands for the space of smooth functions on M with zero mean. Define the subset $\mathcal{H}_{\text{un}} \subset \mathcal{H}$ consisting of all functions $F \in \mathcal{H}$ generating an undistorted one-parameter subgroup, that is $r(F) > 0$. The main result of this section is as follows:

Theorem 6.3.2. *There exists a subset $\mathcal{X} \subset \mathcal{H}_{\text{un}}$ which is C^0-open and C^∞-dense in $\mathcal{H}$.*

Let us mention also that in certain situations (for instance, if $\pi_2(M) = 0$) this result "descends" from the universal cover of $\text{Ham}(M, \omega)$ to the group $\text{Ham}(M, \omega)$ itself. This follows from the discussion on monodromy of Hamiltonian loops in [106].

Finally, similar questions on distortion of *cyclic* subgroups $\{f^n\}_{n \in \mathbb{Z}}$, where f lies in $\text{Ham}(M, \omega)$ or its universal cover, are largely open.

Proof. Denote by $\zeta\colon C^\infty(M) \to \mathbb{R}$ the partial symplectic quasi-state associated to a subadditive spectral invariant c. Inequality (6.1) readily yields

$$|\zeta(F)| \le r(F). \tag{6.3}$$

Consider the subset $\mathcal{X} \subset \mathcal{H}$ consisting of all functions F with $\zeta(F) \neq 0$. By (6.3), $\mathcal{X}$ is contained in $\mathcal{H}_{\text{un}}$. Since ζ is Lipschitz with respect to the uniform norm on $\mathcal{H}$, the set $\mathcal{X}$ is C^0-open in $\mathcal{H}$.

Let us show that $\mathcal{X}$ is dense in $\mathcal{H}$ in the C^∞-topology: Take any open subset $\mathcal{U} \subset \mathcal{H}$. We have to prove that $\mathcal{U} \cap \mathcal{X} \neq \varnothing$. The set $\mathcal{U}$ necessarily contains a Morse function, say F, which attains its maximal value, say a, at exactly one point. If $\zeta(F) \neq 0$, we have that $F \in \mathcal{X}$ and we are done. Otherwise, assume that $\zeta(F) = 0$.

Choose $\delta > 0$ small enough so that $B := f^{-1}\big((a - \delta, a]\big)$ is an open *displaceable* ball in M: there exists $\psi \in \text{Ham}(M, \omega)$ with

$$\psi(B) \cap \text{Closure}(B) = \varnothing.$$

Take any function $u\colon (-\infty, a] \to [0, +\infty)$ which vanishes outside $(a - \delta, a]$ and with $u(a) > 0$. Put $G = u \circ F$. Note that $\langle G \rangle > 0$. Define the following deformation of the function F:

$$F_\epsilon := F + \epsilon \cdot (\langle G \rangle - G), \quad \epsilon \ge 0.$$

Observe that F_ϵ has zero mean. Since $\{F, G\} = 0$ and G is supported in a displaceable set,

$$\zeta(F_\epsilon) = \zeta(F) + \epsilon \cdot \langle G \rangle = \epsilon \cdot \langle G \rangle > 0 \quad \forall \epsilon > 0,$$

and hence $F_\epsilon \in \mathcal{X}$ for all $\epsilon > 0$. Choosing $\epsilon > 0$ so small that $F_\epsilon \in \mathcal{U}$, we get that $\mathcal{U} \cap \mathcal{X} \neq \varnothing$. This completes the proof. □

6.3.2. Hofer's geometry of curves on surfaces. An *equator* is a simple closed curve on S^2 which divides it into two parts of equal area. Denote by $\mathcal{L}$ the set of all equators in S^2. The group $\widetilde{\mathrm{Ham}}(S^2)$ acts on $\mathcal{L}$, and by Exercise 1.3.23 this action is transitive. For $L_1, L_2 \in \mathcal{L}$, set

$$R(L_1, L_2) = \inf\{d(\mathbb{1}, \varphi) : \varphi \in \widetilde{\mathrm{Ham}}(S^2),\ \varphi L_1 = L_2\}. \tag{6.4}$$

It can be shown (see [47]) that R is a metric on $\mathcal{L}$.

Question 2. Is $\mathrm{diam}(\mathcal{L}, R)$ finite or infinite?

This question is still open since the beginning of the 1990s. However, recently M. Khanevsky settled an analogous problem: Consider the open unit disc $D = \{p^2 + q^2 < 1\} \subset \mathbb{R}^2$ equipped with the standard area form $\omega_D = (1/\pi) \cdot dp \wedge dq$. Denote $L_0 = \{q = 0\}$ and define a *diameter* as an embedded interval which divides the disc into two parts of equal area and coincides with L_0 outside a compact set. Denote by $\mathcal{L}$ the set of all diameters, and define a metric R on $\mathcal{L}$ similarly to (6.4). Khanevsky [92] proved that in this case the set of diameters is unbounded:

Theorem 6.3.3. $\mathrm{diam}(\mathcal{L}, R) = \infty$.

A topological remark is in order. The group $\mathrm{Ham}(D)$ is contractible and hence $\widetilde{\mathrm{Ham}}(D) = \mathrm{Ham}(D)$. In what follows we omit tildes.

The proof of Khanevsky's theorem is based on the construction of a Calabi quasi-morphism on the group $\mathrm{Ham}(D)$ presented in [24] (recall that Calabi quasi-morphisms for closed symplectic manifolds were defined in Section 4.8 above).

We start with the unit sphere

$$S^2 = \{x^2 + y^2 + z^2 = 1\} \subset \mathbb{R}^3$$

equipped with the standard symplectic form ω_{S^2}, normalized in such a way that

$$\int_{S^2} \omega_{S^2} = \int_D \omega_D = 1.$$

By Proposition 5.1.10, the two-sphere S^2 admits a subadditive spectral invariant with bounded spectral pseudo-norm which gives rise to the median symplectic quasi-state. Denote by $\mu\colon \widetilde{\mathrm{Ham}}(S^2, \omega) \to \mathbb{R}$ the corresponding Calabi quasi-morphism. Fix $s \in (\frac{1}{2}, 1)$ and define a conformally symplectic embedding $\psi\colon D \to S^2$ such that $\psi^* \omega_{S^2} = s\omega_D$ by mapping the disc onto the cap $\{z + 1 < 2s\} \subset S^2$. The map ψ induces an embedding $\Psi\colon \mathrm{Ham}(D) \to \widetilde{\mathrm{Ham}}(S^2)$. Define

$$\mu_s = \Psi^* \mu\colon \mathrm{Ham}(D) \to \mathbb{R}.$$

The Calabi property of μ implies that for any $f \in \mathrm{Ham}(D)$ generated by a Hamiltonian F_t with $\psi(\bigcup \mathrm{supp}\, F_t)$ displaceable in S^2, $\mu_s(f) = a(s)\,\mathrm{Cal}(f)$, for some constant $a(s)$, where Cal denotes the Calabi homomorphism on $\mathrm{Ham}(D)$.

Exercise 6.3.4. Prove that $a(s) = s^2$.

Exercise 6.3.5. Prove that the quasi-morphisms $\{\mu_s\}_{s \in (1/2,1)}$ are linearly independent.

Next, set $\nu = \mu_s - a(s)\,\mathrm{Cal}$. Then ν is a homogeneous quasi-morphism on $\mathrm{Ham}(D)$. Note that, by construction, $\nu(\varphi) = 0$ whenever $\varphi \in \widetilde{\mathrm{Ham}}(U)$, where $U \subset D$ is a disc of area $< 1/(2s)$. Additionally, the stability of μ implies that there exists $K(s) > 0$ such that

$$|\nu(f) - \nu(g)| \le K \cdot d(f, g) \quad \forall g, f \in \mathrm{Ham}(D).$$

Proof of Theorem 6.3.3. First we claim that ν is bounded on the stabilizer of L_0 in $\mathrm{Ham}(D)$. Indeed, let $f \in \mathrm{Ham}(D)$ such that $fL_0 = L_0$. Since ν is continuous in Hofer's norm, we can perturb f in Hofer's norm, and assume that $f = \mathbb{1}$ in a neighborhood of L_0. Therefore, f decomposes as $f = f_+ f_-$, with f_+, f_- supported in the upper and lower open half-discs, respectively (see Figure 6.3).

Since these half-discs are displaceable, we have $\nu(f_\pm) = 0$, and hence

$$|\nu(f)| = |\nu(f_+ f_-) - \nu(f_+) - \nu(f_+)| \le \Delta,$$

where Δ is the defect of ν. This proves the claim.

Now, take any $g \in \mathrm{Ham}(D)$. We claim that $R(L_0, gL_0) \to \infty$ as $\nu(g) \to \infty$. Indeed, let $f \in \mathrm{Ham}(D)$ such that $fL_0 = gL_0$. Then $f^{-1} g L_0 = L_0$, and hence $|\nu(f^{-1} g)| \le \Delta$. Therefore,

$$\nu(f) \ge \nu(g) + \nu(fg^{-1}) - \Delta \ge \nu(g) - 2\Delta.$$

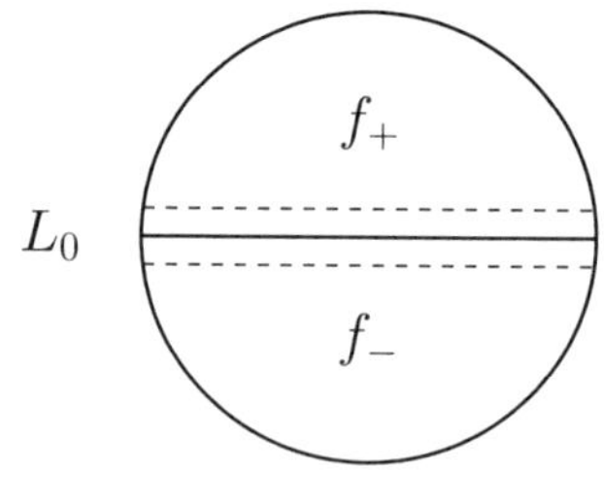

Figure 6.3. Representing f as $f = f_+ f_-$

On the other hand, $\nu(f) \le K \cdot d(\mathbb{1}, f)$. We get

$$d(\mathbb{1}, f) \ge \frac{\nu(g) - 2\Delta}{K}.$$

Taking the infimum over all $f \in \mathrm{Ham}(D)$ such that $fL_0 = gL_0$ we get

$$R(L_0, gL_0) \ge \frac{\nu(g) - 2\Delta}{K}.$$

Thus it remains to find $g_n \in \mathrm{Ham}(D)$ such that $\nu(g_n) \to \infty$ as $n \to \infty$. Noting that ν is a homogeneous quasi-morphism, it suffices to find $g \in \mathrm{Ham}(D)$ such that $\nu(g) \neq 0$. To this end, let G be an autonomous Hamiltonian on D with compact support, and consider the function

$$F = s \cdot G \circ \psi^{-1} \colon \psi(D) \to \mathbb{R}, \tag{6.5}$$

which extends to a smooth function on S^2. Let $g \in \mathrm{Ham}(D)$ denote the time-one map of the Hamiltonian flow of G, and let $\tilde{f} \in \widetilde{\mathrm{Ham}}(S^2)$ be the element represented by the Hamiltonian flow of F. Note that $\tilde{f} = \Psi(g)$, and that

$$\mathrm{Cal}(g) = \int_D G \cdot \omega_D = \int_D s^{-1}\psi^* F \cdot s^{-1}\psi^*\omega_{S^2} = s^{-2}\int_{S^2} F \cdot \omega_{S^2}.$$

Therefore, using Exercise 6.3.4 we get

$$\nu(g) = \mu_s(g) - a(s)\,\mathrm{Cal}(g) = \mu(\tilde{f}) - \int_{S^2} F \cdot \omega_{S^2} = -\zeta(F),$$

where ζ is the median quasi-state. Thus it is enough to find a compactly supported $G \in C^\infty(D)$ such that F, as defined by (6.5), has $\zeta(F) \neq 0$. Such a function clearly exists since $\psi(D)$ is a disc of area $s > \frac{1}{2}$, and hence it contains some equator of the sphere. The proof is complete. □

CHAPTER 7

A Poisson Bracket Invariant of Quadruples

In this chapter, which closely follows [34], we describe a symplectic invariant pb_4 of quadruples of compact subsets of a symplectic manifold. It is defined through an elementary-looking variational problem involving Poisson brackets. The proof of the nontriviality of these invariants involves methods of "hard" symplectic topology. We present applications of pb_4 to Hamiltonian dynamics as well as (in forthcoming chapters) to approximation theory on symplectic manifolds and to quantum mechanics.

7.1. An invariant of quadruples

Let X_0, X_1, Y_0, Y_1 be a quadruple of compact subsets of a symplectic manifold (M, ω). Define the Poisson bracket invariant

$$\mathrm{pb}_4(X_0, X_1, Y_0, Y_1) = \inf \|\{F, G\}\|,$$

where the infimum is taken over the class

$$\begin{aligned}(7.1)\quad & \mathcal{F}_4(X_0, X_1, Y_0, Y_1) \\ & := \{(F, G) : F|_{X_0} \leq 0, F|_{X_1} \geq 1, G|_{Y_0} \leq 0, G|_{Y_1} \geq 1\}\end{aligned}$$

of pairs of functions $F, G \in C_c^\infty(M)$. Note that this class is nonempty whenever

$$X_0 \cap X_1 = Y_0 \cap Y_1 = \varnothing, \tag{7.2}$$

see Figure 7.1.

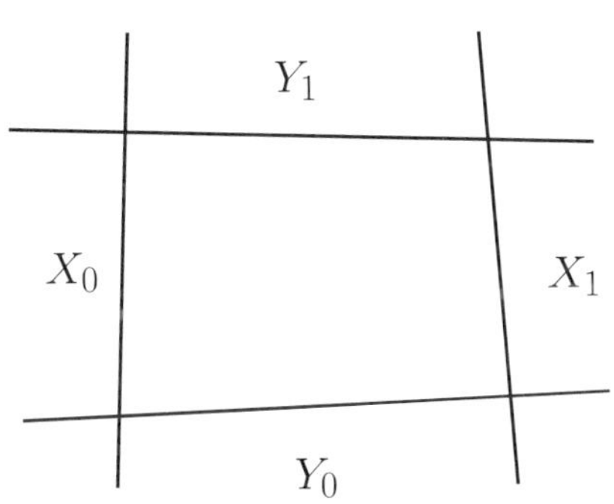

FIGURE 7.1. $X_0 \cap X_1 = Y_0 \cap Y_1 = \varnothing$

If condition (7.2) is violated, we put $\mathrm{pb}_4(X_0, X_1, Y_0, Y_1) = +\infty$. One can easily check that $\mathrm{pb}_4(X_0, X_1, Y_0, Y_1)$ does not change under permutations which switch X_0 with X_1, Y_0 with Y_1, and the pair (X_0, X_1) with the pair (Y_0, Y_1).

In what follows we shall often use a slightly different but equivalent definition of the Poisson bracket invariant. Given a closed subset $X \subset M$, we denote by $\mathrm{Op}(X)$ a sufficiently small neighborhood of X. When we say that $F = 0$ on $\mathrm{Op}(X)$ we mean that F vanishes on *some* neighborhood of X. For a quadruple of compact sets $X_0, X_1, Y_0, Y_1 \subset M$ satisfying (7.2) put

$$\begin{aligned}\mathcal{F}_4'(X_0, X_1, Y_0, Y_1) := \{(F,G) : 0 \leq F \leq 1,\ F|_{\mathrm{Op}(X_0)} = 0,\ F|_{\mathrm{Op}(X_1)} = 1, \\ 0 \leq G \leq 1,\ G|_{\mathrm{Op}(Y_0)} = 0,\ G|_{\mathrm{Op}(Y_1)} = 1\ \}.\end{aligned}$$

Proposition 7.1.1.

$$\mathrm{pb}_4(X_0, X_1, Y_0, Y_1) = \inf_{(F,G)\in\mathcal{F}_4'(X_0,X_1,Y_0,Y_1)} \|\{F,G\}\|. \tag{7.3}$$

PROOF. Set

$$\begin{aligned}\mathrm{pb}_4 &:= \mathrm{pb}_4(X_0, X_1, Y_0, Y_1),\\ \mathrm{pb}_4' &:= \inf_{(F,G)\in\mathcal{F}_4'(X_0,X_1,Y_0,Y_1)} \|\{F,G\}\|.\end{aligned}$$

Clearly,

$$\mathrm{pb}_4 \leq \mathrm{pb}_4',$$

so it suffices to prove that

$$\mathrm{pb}_4 \geq \mathrm{pb}_4'. \tag{7.4}$$

Fix $\epsilon > 0$. Choose $(F, G) \in \mathcal{F}_4(X_0, X_1, Y_0, Y_1)$ so that

$$\mathrm{pb}_4 \geq \|\{F,G\}\| - \epsilon.$$

Fix a small enough $\delta > 0$ and choose a smooth nondecreasing function $u\colon \mathbb{R} \to [0,1]$ so that $u(s) = 0$ for $s \leq \delta$, $u(s) = 1$ for $s \geq 1-\delta$ and $u'(s) \leq 1 + 3\delta$. Put $\phi = u \circ F$ and $\psi = u \circ G$. An immediate check shows that $(\phi, \psi) \in \mathcal{F}_4'(X_0, X_1, Y_0, Y_1)$. Now note that

$$\mathrm{pb}_4' \leq \|\{\phi,\psi\}\| \leq (1+3\delta)^2 \cdot \|\{F,G\}\| \leq (1+3\delta)^2(\mathrm{pb}_4 + \epsilon).$$

Choosing ϵ and δ arbitrarily small, we get (7.4), which completes the proof. □

In the rest of the chapter we develop some methods for establishing lower bounds (and, in particular, the positivity) for the Poisson bracket invariant of certain quadruples of compact subsets of a symplectic manifold.

7.2. Basic properties of pb_4

Let $\mathcal{X} = (X_1, X_2, X_3, X_4)$ be an ordered collection of four compact subsets of a symplectic manifold (M, ω). In what follows $\mathrm{pb}_4(\mathcal{X})$ stands for $\mathrm{pb}_4(X_1, X_2, X_3, X_4)$. Furthermore, we write $\mathrm{pb}_4(\mathcal{X}; M)$ whenever we wish to emphasize the dependence of the Poisson bracket invariant on the ambient symplectic manifold M.

Let $\mathcal{X}$ and $\mathcal{Y}$ be two collections as above. We say that $\mathcal{X} \subset \mathcal{Y}$ if $X_i \subset Y_i$ for all i. Given a compact subset K of a manifold N, we put

$$\mathcal{X} \times K := (X_1 \times K, X_2 \times K, X_3 \times K, X_4 \times K).$$

Let us say that a sequence of subsets $X^{(j)}$ of M *converges* to a limit set Y if every open neighborhood of Y contains all but a finite number of sets from the sequence. This is denoted by $X^{(j)} \to Y$. Given collections $\mathcal{X}^{(j)}$ and $\mathcal{Y}$, we write $\mathcal{X}^{(j)} \to \mathcal{Y}$ if $X_i^{(j)} \to Y_i$ for all $i = 1, \ldots, 4$.

All the properties listed below readily follow from the definitions and Proposition 7.1.1 and are left as exercises.

Semicontinuity. Suppose that a sequence of quadruples $\mathcal{X}^{(j)}$ of subsets of a symplectic manifold converges to a quadruple $\mathcal{Y}$. Then

$$\limsup_{j \to +\infty} \mathrm{pb}_4(\mathcal{X}^{(j)}) \le \mathrm{pb}_4(\mathcal{Y}). \tag{7.5}$$

Behavior under symplectic embeddings. Assume that (M, ω) and (N, σ) are symplectic manifolds of the same dimension. Let $\phi\colon M \to N$ be a symplectic embedding. Let $\mathcal{X}$ and $\mathcal{Y}$ be quadruples of subsets of M and N, respectively, with $\phi(\mathcal{X}) \supset \mathcal{Y}$. Then

$$\mathrm{pb}_4(\mathcal{X}; M) \ge \mathrm{pb}_4(\mathcal{Y}; N). \tag{7.6}$$

In particular, if $\mathcal{X}, \mathcal{Y}$ are quadruples of subsets of M, then

$$\mathcal{X} \supset \mathcal{Y} \implies \mathrm{pb}_4(\mathcal{X}) \ge \mathrm{pb}_4(\mathcal{Y}). \tag{7.7}$$

Behavior under products. Suppose that M and N are connected symplectic manifolds. Equip $M \times N$ with the product symplectic form. Let $K \subset N$ be a compact subset. Then for every collection $\mathcal{X}$ of four compact subsets of M

$$\mathrm{pb}_4(\mathcal{X}, M) \ge \mathrm{pb}_4(\mathcal{X} \times K, M \times N). \tag{7.8}$$

7.3. pb_4 and symplectic quasi-states

Our first bound on pb_4 comes from the theory of symplectic quasi-states. In this section we assume that (M^{2n}, ω) is a closed connected symplectic manifold. Let ζ be the symplectic quasi-state associated to a subadditive spectral invariant with bounded spectral pseudo-norm. Write C for the diameter of $\widetilde{\mathrm{Ham}}(M, \omega)$ with respect to the spectral pseudo-norm.

Theorem 7.3.1. *Let $X_0, X_1, Y_0, Y_1 \subset M$ be a quadruple of closed subsets such that*

$$X_0 \cap X_1 = Y_0 \cap Y_1 = \varnothing$$

and $X_0 \cup Y_0, X_0 \cup Y_1, X_1 \cup Y_0, X_1 \cup Y_1$ are all superheavy. Then

$$\mathrm{pb}_4(X_0, X_1, Y_0, Y_1) \geq \tfrac{1}{8} \cdot C^{-1}. \tag{7.9}$$

We thank Lev Buhovsky and Karina Samvelyan for improving the numerical constant in inequality (7.9).

PROOF. By Proposition 7.1.1, it suffices to find a lower bound on $\|\{F, G\}\|$ for pairs $(F, G) \in \mathcal{F}_4'(X_0, X_1, Y_0, Y_1)$. Put

$$u_1 = FG, \quad u_2 = G(1-F), \quad u_3 = (1-F)(1-G), \quad u_4 = F(1-G).$$

These functions vanish on the superheavy sets $X_0 \cup Y_0, Y_0 \cup X_1, X_1 \cup Y_1, Y_1 \cup X_0$, respectively, and hence $\zeta(u_i) = 0$ for all i. On the other hand, $\sum u_i = 1$ and, as one can easily check,

$$\max_{i,j} \|\{u_i, u_j\}\| \leq \|\{F, G\}\|,$$

and moreover $\{u_1 + u_2, u_3 + u_4\} \equiv 0$. Together with the Poisson bracket inequality (4.16) this yields

$$\begin{aligned} 1 = \zeta\Big(\sum u_i\Big) &= \zeta(u_1 + u_2) + \zeta(u_3 + u_4) \\ &\leq \zeta(u_1) + \zeta(u_2) + \sqrt{2C\|\{u_1, u_2\}\|} + \zeta(u_3) + \zeta(u_4) + \sqrt{2C\|\{u_3, u_4\}\|} \\ &\leq 2 \cdot \sqrt{2C\|\{F, G\}\|}. \end{aligned}$$

Thus $\|\{F, G\}\| \geq \frac{1}{8} \cdot C^{-1}$, as required. □

Example 7.3.2. Let $M = S^2 \times \cdots \times S^2$ be the product of n copies of S^2 equipped with the split symplectic structure $\omega = \sigma \oplus \cdots \oplus \sigma$, where $\int_{S^2} \sigma = 1$. Denote by (x_i, y_i, z_i) the Euclidean coordinates and by (z_i, ϕ_i) the cylindrical coordinates on the ith copy of the sphere ($i = 1, \ldots, n$), where ϕ_i is the polar angle in the (x_i, y_i)-plane. Define the following subsets in the ith factor (see Figure 7.2): Fix $a \in (0, \frac{1}{2})$ so that the σ-area of the set $B_i = \{|z_i| \geq a\}$ is greater than $\frac{1}{2}$. Define the annulus $A_i = \{|z_i| \leq a\}$. Write E_i for the equator $\{z_i = 0\}$ and Q_i^θ for the segment

$$\{\phi_i = \theta\} \cap A_i.$$

Define the following subsets of M:

$$Y_0 = M \setminus \prod \mathrm{Interior}(A_i), \quad Y_1 = \prod E_i.$$

For $v = (\theta_1, \ldots, \theta_n) \in \mathbb{R}^n / 2\pi\mathbb{Z}^n$ denote $Q^v = \prod Q_i^{\theta_i}$. Put $X_0 = Q^v$, $X_1 = Q^w$, where v, w are two distinct points in $\mathbb{R}^n / 2\pi\mathbb{Z}^n$.

We claim that the quadruple X_0, Y_0, X_1, Y_1 satisfies the assumptions of Theorem 7.3.1. The Clifford torus $Y_1 \subset M$ is a stem (see Example 6.1.7),

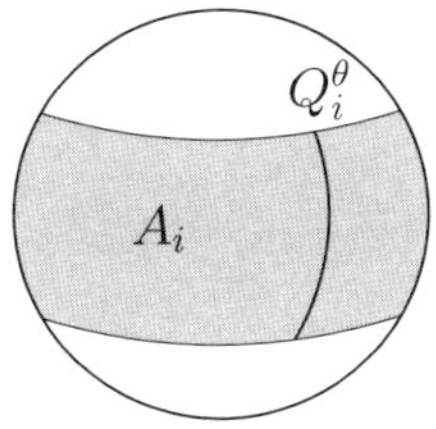

FIGURE 7.2. The subsets A_i and Q_i^θ of the sphere.

and thus it is superheavy by Proposition 6.1.12. Therefore the sets $Y_1 \cup X_i$ are superheavy as well.

Let us check that $X_0 \cup Y_0$ is superheavy. Note that

$$\prod (B_i \cup Q_i^{\theta_i}) \subset Y_0 \cup X_0.$$

But $B_i \cup Q_i^\theta$ is the complement to an open disc of area $< \frac{1}{2}$ and hence is a stem in S^2 (see Example 6.1.3). Since the product of stems is again a stem (Exercise 6.1.4), we conclude that $X_0 \cup Y_0$ is a stem and hence superheavy. Analogously, $X_1 \cup Y_0$ is superheavy, and the claim follows. Thus, $\mathrm{pb}_4(X_0, X_1, Y_0, Y_1) \geq \frac{1}{8}C^{-1}$.

7.4. A dynamical interpretation of pb_4

Let g_t be a Hamiltonian flow on M generated by an autonomous Hamiltonian $G \in C_c^\infty(M)$. We shall refer to pieces of trajectories $\{g_t x\}$, $x \in M$, $t \in [t_0, t_1]$, as *Hamiltonian chords* of G of time-length $t_1 - t_0$. Given two disjoint compact subsets X_0, X_1 of M and a Hamiltonian $G \in C_c^\infty(M)$, denote by $T(X_0, X_1; G)$ the minimal time-length of a Hamiltonian chord of G which connects X_0 and X_1. (Here we set $\inf \varnothing := +\infty$.) Further, set

$$P(X_0, X_1; G) := \inf \|\{F, G\}\|,$$

where the infimum is taken over all functions $F \in C_c^\infty(M)$ with $F|_{X_0} \leq 0, F|_{X_1} \geq 1$.

Theorem 7.4.1.

$$T(X_0, X_1; G) = P(X_0, X_1; G)^{-1}.$$

As an immediate consequence of Theorem 7.4.1, we get the following interpretation of the Poisson bracket invariant pb_4. Let X_0, X_1, Y_0, Y_1 be a quadruple of compact subsets satisfying intersection condition (7.2). Put

$$T(X_0, X_1, Y_0, Y_1) = \sup\{\, T(X_0, X_1; G) : G \in C_c^\infty(M),\ G|_{Y_0} \leq 0,\ G|_{Y_1} \geq 1\,\}.$$

Corollary 7.4.2.

$$\mathrm{pb}_4(X_0, X_1, Y_0, Y_1) = T(X_0, X_1, Y_0, Y_1)^{-1}.$$

PROOF. Indeed,

$\mathrm{pb}_4(X_0, X_1, Y_0, Y_1) = \inf\{P(X_0, X_1; G) : G \in C_c^\infty(M), G|_{Y_0} \le 0, G|_{Y_1} \ge 1\}$, and now result follows from Theorem 7.4.1. □

For the proof of Theorem 7.4.1 we need the following auxiliary result.

Lemma 7.4.3. *Let v be a smooth compactly supported vector field on M and let X_0, X_1 be a pair of disjoint compact subsets of M. Denote by g_t the flow of v. Assume that $g_t X_0 \cap X_1 = \varnothing$ for all $t \in [-a, a]$ for some $a > 0$. Then there exists a smooth compactly supported function $F\colon M \to [0,1]$ such that $F|_{X_0} = 0$, $F|_{X_1} = 1$ and the Lie derivative $\mathcal{L}_v F$ satisfies $\|\mathcal{L}_v F\| < 1/a$.*

PROOF. Choose $b > a$, sufficiently close to a, so that $g_t X_0 \cap X_1 = \varnothing$ for all $t \in [-b, b]$. The sets

$$\check{X}_0 := \bigcup_{t\in[0,b]} g_t(X_0)$$

and

$$\check{X}_1 := \bigcup_{t\in[0,b]} g_t(X_1)$$

do not intersect. Take any smooth compactly supported function $H\colon M \to [0,1]$ such that $H = 0$ on $\check{X}_0$ and $H = 1$ on $\check{X}_1$. Put

$$F := \frac{1}{b}\int_0^b H \circ g_t \,\mathrm{d}t.$$

Clearly, F takes values in $[0,1]$, is compactly supported, $F = 0$ on X_0, $F = 1$ on X_1 and

$$\mathcal{L}_v F = \frac{1}{b}\int_0^b \frac{\mathrm{d}}{\mathrm{d}t} H \circ g_t \,\mathrm{d}t = \frac{1}{b}(H \circ g_b - H).$$

It follows that $\|\mathcal{L}_v F\| \le 1/b < 1/a$, and we are done. □

PROOF OF THEOREM 7.4.1. Put $T := T(X_0, X_1; G)$ and let F be any smooth compactly supported function with

$$F|_{X_0} \le 0, \quad F|_{X_1} \ge 1.$$

There exist $i \in \{0,1\}$ and $x \in X_i$ so that $g_T x \in X_{1-i}$ (recall that g_t is the Hamiltonian flow of G). Thus $|F(g_T x) - F(x)| \ge 1$, which yields $\|\{F, G\}\| \ge T^{-1}$. Therefore

$$T \ge \frac{1}{P(X_0, X_1; G)}.$$

It remains to prove the opposite inequality. Put $v = \operatorname{sgrad} G$. Assume, on the contrary, that

$$T > \frac{1}{P(X_0, X_1; G)}.$$

Thus there exists

$$a > \frac{1}{P(X_0, X_1; G)}$$

so that $g_t X_0 \cap X_1 = \varnothing$ for all $t \in [-a, a]$. By Lemma 7.4.3, there exists a smooth compactly supported function $F\colon M \to [0, 1]$ such that $F|_{X_0} = 0$, $F|_{X_1} = 1$ and $\|\mathcal{L}_v F\| < 1/a$. But $\mathcal{L}_v F = \{F, G\}$ and we conclude that $P(X_0, X_1; G) < 1/a$, a contradiction. This completes the proof. □

Let us spell out the following (tautological) consequence of Corollary 7.4.2:

Theorem 7.4.4. *Let $X_0, X_1, Y_0, Y_1 \subset M$ be a quadruple of compact sets with $X_0 \cap X_1 = Y_0 \cap Y_1 = \varnothing$ and $\mathrm{pb}_4(X_0, X_1, Y_0, Y_1) = p > 0$. Let $G \in C_c^\infty(M)$ be a Hamiltonian function with $G|_{Y_0} \le 0$ and $G|_{Y_1} \ge 1$ generating a Hamiltonian flow g_t. Then $g_T x \in X_1$ for some point $x \in X_0$ and some time moment $T \in [-1/p, 1/p]$.*

This result has a number of applications. First, consider a compact domain $V \subset T^*\mathbb{T}^n$ whose interior contains the zero-section $\mathbb{T}^n$. Fix a pair of distinct points $q_0, q_1 \in \mathbb{T}^n$ and put $D_i = T^*_{q_i}\mathbb{T}^n \cap V$, $i = 0, 1$.

Theorem 7.4.5. *Let $G\colon V \to \mathbb{R}$ be a Hamiltonian which vanishes near ∂V and such that $G \ge 1$ on the zero section. Then there exists a Hamiltonian chord of the Hamiltonian flow of G passing through D_0 and D_1.*

As an illustration, assume that the torus $\mathbb{T}^n$ is equipped with a Riemannian metric and $V = \{|p| \le 1\}$, where (p, q) are canonical coordinates on $T^*\mathbb{T}^n$. Suppose that the Hamiltonian G has the form $u(|p|)$, where $u(s)$ vanishes for s close to 1 and $u(0) = 1$. Then the projection of the Hamiltonian chord provided by Theorem 7.4.5 is a (reparameterized) Riemannian geodesic segment joining the points q_0 and q_1.

We refer to [34] for a generalization of Theorem 7.4.5 to nonautonomous Hamiltonian flows.

PROOF OF THEOREM 7.4.5. Write $T^*\mathbb{T}^n$ as the n-fold product $T^*S^1 \times \cdots \times T^*S^1$. It is convenient to consider S^1 as $\mathbb{R}/2\pi\mathbb{Z}$. Let (z_i, ϕ_i) be the canonical coordinates on the ith factor $T^*S^1 = \mathbb{R} \times S^1$. The symplectic form on $T^*\mathbb{T}^n$ in these coordinates is given by $\sum_i \sigma_i$, where $\sigma_i = \frac{1}{2\pi} \mathrm{d}z_i \wedge \mathrm{d}\phi_i$. Fix $a \in (0, \frac{1}{2})$. Define, in the ith copy of T^*S^1, the annulus $A_i = \{|z_i| \le a\}$. Making, if necessary, the rescaling $(p, q) \to (\mu p, q)$ with a sufficiently small $\mu > 0$, we can assume that the domain V from the formulation of the theorem is contained in the interior of $\prod_i A_i$.

Compactify now the subset $\prod_i \{|z_i| < 1\} \subset T^*\mathbb{T}^n$ into the closed symplectic manifold $M = S^2 \times \cdots \times S^2$ equipped with the split symplectic structure $\sigma \oplus \cdots \oplus \sigma$. After the compactification, (z_i, ϕ_i) become the cylindrical coordinates on the i-th factor S^2 of M. Let us consider the Hamiltonian G from the formulation of the theorem as a function on M. We extend it by zero to the complement of V.

At this point we find ourselves in the setting of Example 7.3.2 above. Let X_0, X_1, Y_0, Y_1 be the quadruple of subsets from this example. Recall that

- $Y_0 = M \setminus \prod_i \mathrm{Interior}(A_i)$;

- Y_1 is the torus $\prod_i \{z_i = 0\}$, which in the original picture corresponds to the zero section;
- X_0 and X_1 are discs corresponding to the intersection of $\prod_i A_i$ with two distinct cotangent fibers.

Without loss of generality assume that the sets D_0 and D_1 from the formulation of the theorem are identified with $X_0 \cap V$ and $X_1 \cap V$, respectively.

According to Example 7.3.2,

$$\mathrm{pb}_4(X_0, X_1, Y_0, Y_1) > 0.$$

Thus by Theorem 7.4.4 the Hamiltonian flow g_t of G has a chord passing through X_0 and X_1. Since g_t is the identity outside V, this chord is entirely contained in V and thus it connects D_0 and D_1. This completes the proof. □

As a second application of Theorem 7.4.4 we prove the following vanishing result for pb_4 which will be used below. Let us introduce the following terminology: Assume that S is a finite simplicial complex and M is a manifold. Let $\phi\colon S \to M$ be a homeomorphism from S to its image $\phi(S)$ such that the restriction of ϕ to every simplex is a smooth embedding. We refer to the image $X := \phi(S)$ as an embedded simplicial complex in M. We denote by $\dim X$ the maximal dimension of a simplex from S.

Proposition 7.4.6. *Let X_0, X_1 be disjoint embedded simplicial complexes in a symplectic manifold (M, ω). Assume that*

$$\dim X_0 + \dim X_1 \leq 2n - 2. \tag{7.10}$$

Then for every pair of disjoint compact subsets Y_0, Y_1 of M,

$$\mathrm{pb}_4(X_0, X_1, Y_0, Y_1) = 0.$$

PROOF. Assume, on the contrary, that $\mathrm{pb}_4(X_0, X_1, Y_0, Y_1) = p > 0$. Fix disjoint neighborhoods U_i of Y_i, $i = 0, 1$. Take any function $H \in C_c^\infty(M)$ such that $H = 0$ on U_0 and $H = 1$ on U_1. Denote by h_t the Hamiltonian flow generated by H. Put $T = 2/p$ and set $Z := \bigcup_{|t| \leq T} h_t(X_0)$. The dimension formula (7.10) and a standard transversality argument yield the existence of an arbitrary C^0-small Hamiltonian diffeomorphism f of M supported near X_1 such that

$$f(X_1) \cap Z = \varnothing. \tag{7.11}$$

Since f is C^0-small, we can assume that $f(Y_i) \subset U_i$ for $i = 0, 1$. Moreover, since f is supported near X_1 and $X_0 \cap X_1 = \varnothing$, we have $f(X_0) = X_0$. By the symplectic invariance of pb_4,

$$\mathrm{pb}_4\big(f(X_0), f(X_1), f(Y_0), f(Y_1)\big) = p.$$

Theorem 7.4.4 guarantees that there exists a Hamiltonian chord of h_t of time-length $\leq 1/p$ joining $f(X_0) = X_0$ and $f(X_1)$. This contradicts property (7.11), and hence $p = 0$. □

7.5. pb_4 and deformations of the symplectic form

In this section we present yet another approach to the positivity of the Poisson bracket invariant which is applicable to certain quadruples of (sometimes singular) Lagrangian submanifolds. Our method is based on a special deformation of the symplectic form on M^{2n} combined with the study of "persistent" pseudo-holomorphic curves with Lagrangian boundary conditions (cf. [2]).

7.5.1. Topological preliminaries. Let X_0, X_1, Y_0, Y_1 be a quadruple of compact subsets of M with $X_0 \cap X_1 = Y_0 \cap Y_1 = \varnothing$. Let (F, G) be a pair of functions from $\mathcal{F}'_4(X_0, X_1, Y_0, Y_1)$. Recall that this means that F and G are compactly supported smooth functions $M \to [0,1]$, F vanishes near X_0 and equals 1 near X_1, and G vanishes near Y_0 and equals 1 near Y_1. It follows that the map

$$\Phi\colon M \to \mathbb{R}^2, \quad x \mapsto \big(F(x), G(x)\big),$$

sends the union $W := X_0 \cup Y_0 \cup X_1 \cup Y_1$ to the boundary Γ of the unit square $[0,1] \times [0,1]$. Orient Γ counter-clockwise. We say that the class $e \in H_1(W;\mathbb{Z})$ is *basic* if $\Phi_* e = [\Gamma] \in H_1(\Gamma;\mathbb{Z})$. Further, observe that the 1-form $F\,\mathrm{d}G$ is necessarily closed in a sufficiently small neighborhood U of W.

Exercise 7.5.1. (i) Show that the image of $[F\,\mathrm{d}G]$ in $H^1(W;\mathbb{R})$ under the natural morphism $H^1(U;\mathbb{R}) \to H^1(W;\mathbb{R})$ does not depend on the specific choice of (F,G).

(ii) Show that $\int_\gamma F\,\mathrm{d}G = 1$ for every loop $\gamma \subset U$ representing a basic class in $H_1(W;\mathbb{Z})$.

7.5.2. A lower bound. We keep the notation of Section 7.5.1. Given a pair of functions $(F,G) \in \mathcal{F}'_4(X_0, X_1, Y_0, Y_1)$, consider the family of forms

$$\omega_s := \omega - s\,\mathrm{d}F \wedge \mathrm{d}G, \quad s \geq 0.$$

Recall (see Exercise 1.3.12) that

$$\mathrm{d}F \wedge \mathrm{d}G \wedge \omega^{n-1} = -\frac{1}{n}\{F,G\} \cdot \omega^n.$$

Therefore,

$$\omega_s^n = (1 + s\{F,G\})\omega^n.$$

The Poisson bracket $\{F,G\}$ either has zero mean (if M is closed), or vanishes outside a compact subset (if M is open), so necessarily $\min\{F,G\} \leq 0$. It follows that the form ω_s is symplectic if and only if $s < 1/|\min\{F,G\}|$. In particular, it is symplectic for all

$$s \in I := [0, 1/\|\{F,G\}\|).$$

(We set $1/\|\{F,G\}\| = +\infty$ if $\{F,G\} \equiv 0$.)[1]

[1]The above consideration is presented in [34] with inaccurately chosen signs. This fortunately does not affect any results.

The next elementary proposition allows us to relate Poisson brackets to pseudo-holomorphic curves (recall Section 1.6):

Proposition 7.5.2. *Consider a pair* (F, G) *from* $\mathcal{F}_4'(X_0, X_1, Y_0, Y_1)$. *Assume that there exist*

- *a family of almost complex structures* J_s, $s \in I$, *such that each* J_s *is compatible with the symplectic form* $\omega_s = \omega - s\,\mathrm{d}F \wedge \mathrm{d}G$;
- *a family of* J_s*-holomorphic maps* $u_s\colon (\Sigma_s, \partial\Sigma_s) \to (M, W)$, $s \in I$, *where each* Σ_s *is a compact Riemann surface with boundary and possibly with corners, such that* $u_s(\partial\Sigma_s)$ *represents a basic class in* $H_1(W; \mathbb{Z})$;
- *a positive constant* C *such that* $\int_{\Sigma_s} u_s^*\omega \le C$ *for all* $s \in I$.

Then $\|\{F, G\}\| \ge 1/C$.

PROOF. Applying the Stokes theorem together with Exercises 7.5.1(ii) and 1.6.4 we get that

$$0 \le \int_{\Sigma_s} u_s^*\omega_s = \int_{\Sigma_s} u_s^*\omega - s\int_{\partial\Sigma_s} u_s^*(F\,\mathrm{d}G) \le C - s$$

for any $s \in I = [0, 1/\|\{F, G\}\|)$. It follows that $\|\{F, G\}\| \ge 1/C$. □

7.5.3. Quadrilaterals on surfaces. In the rest of this chapter we demonstrate a delicate balance between rigidity and flexibility of the pb_4-invariant. We start with the following simple situation. Let (M, ω) be a symplectic surface of area $B \in (0, +\infty]$. Consider a curvilinear quadrilateral $\Pi \subset M$ of area A with sides denoted in the cyclic order by a_1, a_2, a_3, a_4; i.e., Π is a topological disc bounded by the union of four smooth embedded curves a_1, a_2, a_3, a_4 connecting four distinct points in M in cyclic order as listed here and (transversally) intersecting each other only at their common end-points. Our first objective is to calculate the value of $\mathrm{pb}_4(a_1, a_3, a_2, a_4)$. Recall from Section 7.1 that

$$\mathrm{pb}_4(a_1, a_3, a_2, a_4) = \mathrm{pb}_4(a_1, a_3, a_4, a_2). \tag{7.12}$$

Thus without loss of generality we can assume that the orientation of $\partial\Pi$ induced by the cyclic order of a_i's coincides with the boundary orientation.

Theorem 7.5.3. $\mathrm{pb}_4(a_1, a_3, a_2, a_4) = \max\bigl(1/A, 1/(B - A)\bigr)$.

PROOF.

LOWER BOUND. Pick any $(F, G) \in \mathcal{F}_4'(a_1, a_3, a_2, a_4)$. Note that the quadrilateral Π is J-holomorphic for *any* (almost) complex structure J on M compatible with the orientation. Also note that by a direct calculation $\int_{\partial\Pi} F\,\mathrm{d}G = 1$. Thus one can apply Proposition 7.5.2 with $\Sigma = \Pi$ and get that

$$\|\{F, G\}\| \ge 1/A.$$

Since this is true for any $(F, G) \in \mathcal{F}_4'(a_1, a_3, a_2, a_4)$, we get that

$$\mathrm{pb}_4(a_1, a_3, a_2, a_4) \ge 1/A. \tag{7.13}$$

Further, if M is a closed surface apply Proposition 7.5.2 with $\Sigma = \overline{M \setminus \Pi}$. We get that (mind the order of sides)

$$\mathrm{pb}_4(a_1, a_3, a_4, a_2) \geq 1/(B - A). \tag{7.14}$$

If M is open, the surface Σ is not compact. However, since Σ is properly embedded and the functions F and G are compactly supported, Proposition 7.5.2 is still applicable (after an obvious modification) and yields inequality (7.14).

Combining inequalities (7.13) and (7.14) with (7.12) we get that

$$\mathrm{pb}_4(a_1, a_3, a_2, a_4) \geq \max\big(1/A, 1/(B - A)\big). \tag{7.15}$$

UPPER BOUND.[2] Fix any positive $C \in (A, B)$ and a sufficiently small $\epsilon > 0$. By Moser's theorem [114], we can assume that M contains a rectangle

$$K = [-\epsilon, C + \epsilon] \times [-2\epsilon, 1 + 2\epsilon]$$

equipped with coordinates (p, q) such that the symplectic form ω equals $dp \wedge dq$ and the quadrilateral Π is given by $[0, A] \times [0, 1]$. Let a_1, a_2, a_3, a_4 be the sides of Π taken counter-clockwise starting from $a_1 = \{0\} \times [0, 1]$.

Introduce the following four real-valued functions u_1, v_1, u_2, v_2 on $\mathbb{R}$:

- $u_1(A) = 1$, $\mathrm{supp}(u_1) \subset [0, C]$, $|u_1'| \leq \max\big(1/A, 1/(C - A)\big) + \epsilon$;
- $|v_1| \leq 1$, $v_1 = 1$ on $[0, 1]$, $\mathrm{supp}(v_1) \subset [-\epsilon, 1 + \epsilon]$;
- $|u_2| \leq 1$, $u_2 = 1$ on $[0, C]$, $\mathrm{supp}(u_2) \subset [-\epsilon, C + \epsilon]$;
- $v_2(s) = s$ on $[-\epsilon, 1 + \epsilon]$, $\mathrm{supp}(v_2) \subset [-2\epsilon, 1 + 2\epsilon]$.

Put $F(p, q) = u_1(p)v_1(q)$ and $G(p, q) = u_2(p)v_2(q)$. One readily checks that F and G are supported in K and $(F, G) \in \mathcal{F}_4(a_1, a_3, a_2, a_4)$. Furthermore,

$$\{F, G\} = u_1(p)v_2(q)v_1'(q)u_2'(p) - u_2(p)v_1(q)u_1'(p)v_2'(q). \tag{7.16}$$

Observe that u_2 is constant on $[0, C]$ and $\mathrm{supp}(u_1) \subset [0, C]$. Therefore, $u_1(p)u_2'(p) = 0$, and hence the first term in (7.16) vanishes. Since v_1 is supported in the interval $[-\epsilon, 1 + \epsilon]$ and $v_2' = 1$ on this interval, we get that $v_1(q)v_2'(q) = v_1(q)$ and hence

$$\|\{F, G\}\| = \|u_2(p)v_1(q)u_1'(p)\| \leq \max\big(1/A, 1/(C - A)\big) + \epsilon.$$

Letting $C \to B$ and $\epsilon \to 0$, we conclude that

$$\mathrm{pb}_4(a_1, a_3, a_2, a_4) \leq \max\big(1/A, 1/(B - A)\big).$$

Together with (7.15), this yields the theorem. □

[2]The proof of the upper bound given in [34, p. 105] contains a mistake. The correction presented here is due to Lev Buhovsky.

7.5.4. An effect of stabilization. Stabilization provides an instructive way of testing validity of two-dimensional symplectic phenomena in higher dimensions. By definition, *a stabilization* of a subset $X \subset (M, \omega)$ is a subset of the form

$$X \times K \subset (M \times T^* S^1, \Omega), \quad \Omega = \omega \oplus dp \wedge dq,$$

where $K = \{p = dF/dq\} \subset T^* S^1$ with $F \in C^\infty(S^1)$. Such a K is called an exact Lagrangian section of $T^* S^1$. It turns out that our estimates of the pb_4-invariant for quadrilaterals on surfaces behave in quite a subtle way under stabilization. We use the notation from the previous section. In particular, M is a symplectic surface.

Suppose first that $M \neq S^2$. Let K be an exact Lagrangian section of $T^* S^1$. We claim that

$$\mathrm{pb}_4(a_1 \times K, a_3 \times K, a_2 \times K, a_4 \times K) \geq 1/A. \tag{7.17}$$

Indeed, after a C^0-perturbation $a_1^\epsilon, a_2^\epsilon, a_3^\epsilon, a_4^\epsilon$ of a_1, a_2, a_3, a_4 we can assume that

$$L = a_1^\epsilon \cup a_2^\epsilon \cup a_3^\epsilon \cup a_4^\epsilon$$

is a smooth embedded circle in M enclosing area A^ϵ. Take a split complex structure J on $M \times T^* S^1$. Observe that $\widehat{L} = L \times K$ is a Lagrangian torus in $M \times T^* S^1$. As the deformation parameter s changes, $\widehat{L}$ remains a Lagrangian submanifold for the deformed symplectic structure $\Omega_s = \Omega - s\, dF \wedge dG$, provided

$$(F, G) \in \mathcal{F}_4'(a_1^\epsilon \times K, a_3^\epsilon \times K, a_2^\epsilon \times K, a_4^\epsilon \times K),$$

but its symplectic area class alters. (By Moser's theorem [114], an equivalent viewpoint is that the symplectic form on $M \times T^* S^1$ is fixed, but $\hat{L}$ undergoes the process of a Lagrangian, albeit non-Hamiltonian, isotopy.) The class $\alpha := [\Pi \times \text{point}]$ is the generator of $\pi_2(M \times T^* S^1, L \times K)$. Thus standard Gromov theory [76] yields that for a generic deformation J_s of J as in Proposition 7.5.2 there exists a pseudo-holomorphic disc Σ_s in the class α (this argument breaks down for $M = S^2$ due to possible bubbling). Therefore, Proposition 7.5.2 with $C = A^\epsilon$ yields

$$\mathrm{pb}_4(a_1^\epsilon \times K, a_3^\epsilon \times K, a_2^\epsilon \times K, a_4^\epsilon \times K) \geq 1/A^\epsilon.$$

Passing to the limit as the size of perturbation ϵ goes to 0 (this procedure is justified in (7.5)) we get inequality (7.17).

Let us emphasize that a Gromov-theoretical argument as above does not work for surfaces Σ other than discs, and in particular it is not applicable to $\overline{M \setminus \Pi}$.

Example 7.5.4. Assume that $\mathrm{Area}(M) = 2\,\mathrm{Area}(\Pi)$. Let $W_\epsilon \subset T^* S^1$ be the annulus $\{|p| < \epsilon\}$, $\epsilon > 0$, and let $K \subset W_\epsilon$ be the zero section. Combining Theorem 7.5.3, properties (7.8), (7.6), and inequality (7.17), we get that

$$\frac{1}{A} = \mathrm{pb}_4(a_1, a_3, a_2, a_4; M) \geq \mathrm{pb}_4(a_1 \times K, a_3 \times K, a_2 \times K, a_4 \times K; M \times W_\epsilon)$$
$$\geq \mathrm{pb}_4(a_1 \times K, a_3 \times K, a_2 \times K, a_4 \times K; M \times T^*S^1) \geq \frac{1}{A},$$

and hence

$$\mathrm{pb}_4(a_1 \times K, a_3 \times K, a_2 \times K, a_4 \times K; M \times W_\epsilon) = \frac{1}{A}. \tag{7.18}$$

The proof of (7.17) does not work for $M = S^2$. However in this case one can prove the following stronger result:

Proposition 7.5.5. *Assume that $M = S^2$. Then*

$$\mathrm{pb}_4(a_1 \times K, a_3 \times K, a_2 \times K, a_4 \times K) = \max\bigl(1/A, 1/(B - A)\bigr).$$

The proof is based on a method of symplectic field theory, see [34].

In contrast to the previous situation, if one stabilizes each of the sets a_1, a_2, a_3, a_4 by *its own* exact Lagrangian section of T^*S^1, a transition from rigidity to flexibility takes place:

Proposition 7.5.6. *Let $K_1, \ldots, K_4$ be a generic quadruple of sections of T^*S^1. Then*

$$p := \mathrm{pb}_4(a_1 \times K_1, a_3 \times K_3, a_2 \times K_2, a_4 \times K_4) = 0.$$

PROOF. Consider the annulus $A = [-1, 1] \times S^1$. Let $x_1, \ldots, x_k, y_1, \ldots, y_l$ be pairwise distinct points in S^1. Consider the 1-dimensional simplicial complex

$$A(x_1, \ldots, x_k, y_1, \ldots, y_l) \subset A,$$

defined by

$$A(x_1, \ldots, x_k, y_1, \ldots, y_l) = (\{0\} \times S^1) \cup \bigcup_{i=1}^{k} ([0, 1] \times \{x_i\}) \cup \bigcup_{j=1}^{l} ([-1, 0] \times \{y_j\}).$$

Put $P_i = a_i \times K_i$. We will use the cyclic convention for the indices $i = 1, 2, 3, 4$ (that is $4{+}1 = 1$, $1{-}1 = 4$). Choose in the obvious way parametrizations $\phi_i \colon A \to P_i$, $i = 1, 2, 3, 4$, such that $P_i \cap P_{i+1}$ consists of a finite number of points of the form

$$\phi_i(1, x_j) = \phi_{i+1}(-1, y_j), \quad j = 1, \ldots, N(i),$$

for some $N(i) \in \mathbb{N}$. Put

$$S_i = A\bigl(x_1, \ldots, x_{N(i)}, y_1, \ldots, y_{N(i-1)}\bigr).$$

Fix $\epsilon > 0$ and observe that one can "collapse" the annulus A to an ϵ-neighborhood S_i^ϵ of S_i. More precisely, there exists a family of embeddings $\psi_i^t \colon A \to A$, $t \in [0, 1]$, such that $\psi_i^0 = \mathbb{1}$, $\psi_i^1(A) = S_i^\epsilon$, and $\psi_i^t = \mathbb{1}$ near S_i for all t.

Put $Q_i = \phi_i(S_i)$ and $Q_i^\epsilon = \phi_i(S_i^\epsilon)$. Observe that $P_i \cap P_{i+1} = Q_i \cap Q_{i+1}$ and therefore the isotopies

$$\theta_i^t := \phi_i \circ \psi_i^t \colon A \to P_i$$

have disjoint supports for distinct i. Since each P_i is a Lagrangian submanifold of $M \times T^*S^1$, the isotopies θ_i^t, $i = 1,2,3,4$, extend *simultaneously* to an ambient Hamiltonian isotopy θ^t of $M \times T^*S^1$. By the symplectic invariance of pb_4,

$$\mathrm{pb}_4(Q_1^\epsilon, Q_3^\epsilon, Q_2^\epsilon, Q_4^\epsilon) = p \tag{7.19}$$

for all ϵ. Note that $Q_i^\epsilon \to Q_i$ as $\epsilon \to 0$, where the convergence is understood in the sense of Section 7.2. Thus, by (7.5),

$$p \leq q := \mathrm{pb}_4(Q_1, Q_3, Q_2, Q_4).$$

Since each Q_i is a one-dimensional embedded simplicial complex in $M \times T^*S^1$, Proposition 7.4.6 yields $q = 0$. Hence $p = 0$, which completes the proof. $\square$

Replacing the sides a_1, a_2, a_3, a_4 of the quadrilateral by larger subsets, one can restore rigidity of the pb_4-invariant even when one stabilizes each of the subsets by its own exact Lagrangian section of T^*S^1. As an example, consider the case when M is a symplectic two-torus $\mathbb{T}^2$ of total area 1 and X_0, X_1 and Y_0, Y_1 are pairs of parallels and meridians respectively which divide the torus into four squares. Pick one of them and denote it by Π. Let $a_1 \subset X_0$, $a_2 \subset Y_0$, $a_3 \subset X_1$, $a_4 \subset Y_1$ be the sides of Π. Write A for the area of Π. Theorem 7.5.3 and monotonicity of pb_4 with respect to inclusions of sets (see (7.7) above) yield

$$\mathrm{pb}_4(X_0, X_1, Y_0, Y_1) \geq \mathrm{pb}_4(a_1, a_3, a_2, a_4) \geq 1/A > 0.$$

Furthermore, if K is an exact section of T^*S^1, then by (7.17)

$$\begin{aligned} &\mathrm{pb}_4(X_0 \times K, X_1 \times K, Y_0 \times K, Y_1 \times K) \\ &\qquad \geq \mathrm{pb}_4(a_1 \times K, a_3 \times K, a_2 \times K, a_4 \times K) \geq 1/A. \end{aligned}$$

If K_1, K_2, K_3, K_4 is a generic quadruple of exact Lagrangian sections of T^*S^1, then, according to Proposition 7.5.6,

$$\mathrm{pb}_4(a_1 \times K_1, a_3 \times K_3, a_2 \times K_2, a_4 \times K_4) = 0.$$

However

$$\mathrm{pb}_4(X_0 \times K_1, X_1 \times K_3, Y_0 \times K_2, Y_1 \times K_4) > 0.$$

The proof of the latter claim is based of Proposition 7.5.2 combined with the existence of persistent pseudo-holomorphic squares whose edges lie on

$$\widetilde{X_i} \times S^1, \widetilde{Y_i} \times S^1, \quad i = 0, 1,$$

where $\widetilde{X_i}, \widetilde{Y_i} \subset \mathbb{R}^2$ are lifts of the corresponding circles to the universal cover $\mathbb{R}^2$ of $\mathbb{T}^2$. Here persistent pseudo-holomorphic squares come from the μ^3-operation in the Donaldson–Fukaya category, which lies out of the scope of the present book. We refer to [34] for details.

CHAPTER 8

Symplectic Approximation Theory

The problem of symplectic approximation theory on a symplectic manifold (M, ω) can be roughly formulated as follows: Given a pair of functions F and G on M, find their (optimal) approximation with respect to the uniform norm by a pair of functions with "small" Poisson bracket. This can be considered as a symplectic counterpart of the following classical problem (see [93] for a solution and an extensive discussion): Given a function (say, of one real variable), find its optimal approximation by a function with a given bound on the derivative. The analogy comes from the fact that the Poisson bracket $\{F, G\}$ is the (Lie) derivative $\mathcal{L}_{\mathrm{sgrad}\, G}F$ of F with respect to the Hamiltonian vector field $\mathrm{sgrad}\, G$. The major difference is that in the symplectic case we approximate not only the argument F, but also the differential operator $\mathcal{L}_{\mathrm{sgrad}\, G}$ which depends on the derivatives of G and which may change drastically under C^0-perturbations of G. This is exactly the point where symplectic topology enters the stage. Let us pass to a more precise discussion.

8.1. The profile function

In this section we follow [34]. Let (M, ω) be a symplectic manifold. Denote by $C_c^\infty(M)$ the set of smooth compactly supported functions and set $\mathcal{F} = C_c^\infty(M) \times C_c^\infty(M)$. Equip $\mathcal{F}$ with the metric

$$\rho\big((F, G), (F', G')\big) = \|F - F'\| + \|G - G'\|.$$

Set

$$\mathcal{K}_s = \{(A, B) \in \mathcal{F} : \|\{A, B\}\| \le s\}.$$

Take any pair $(F, G) \in \mathcal{F}$ with $\{F, G\} \neq 0$. Define the *profile function*

$$r(s) = r_{F,G}(s) = \rho((F, G), \mathcal{K}_s).$$

With this language, *the objective of symplectic approximation theory is to study the behavior of the profile function.*

Proposition 8.1.1 (Properties of the profile function).

(1) *r is nonincreasing.*
(2) *$r(s) = 0$ for any $s \ge \|\{F, G\}\|$.*
(3) *$r(s) > 0$ for any $s < \|\{F, G\}\|$.*
(4) *There exists a constant $C(F, G)$ such that*

$$r(\|\{F, G\}\| - s) \ge C(F, G) \cdot s^{3/2}.$$

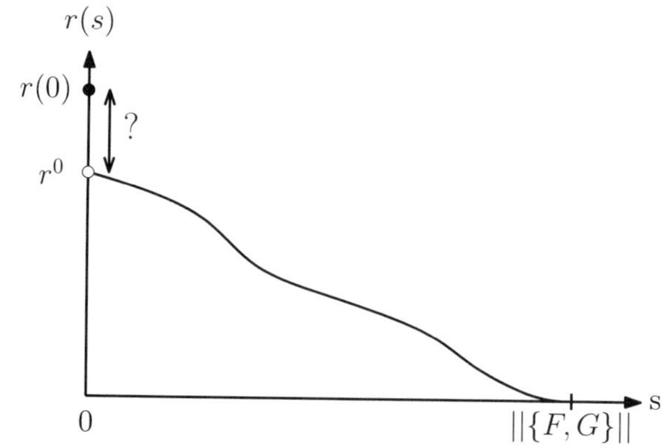

FIGURE 8.1. The profile function

(5) $r \leq \min(\|F\|, \|G\|)$.
(6) r *is continuous outside* 0.

These properties are illustrated in Figure 8.1. Before proving them, we observe that property (1) yields the existence of the limit

$$r^0 := \lim_{s \searrow 0} r(s)$$

and the inequality $r^0 \leq r(0)$. The delicate question of when equality holds is addressed in Section 8.2 below.

PROOF OF PROPOSITION 8.1.1. Properties (1), (2) and (5) immediately follow from the definitions (for property (5), look at the distance from (F, G) to $(F, 0)$ and $(0, G)$).

Properties (3) and (4) are simply reformulations of the lower semi-continuity of the functional $(F, G) \mapsto \|\{F, G\}\|$ (Theorem 2.1.1 above) and of Buhovsky's 2/3 law (Theorem 2.1.7 above), respectively. Thus the profile function provides a more general context for the study of C^0-rigidity of the Poisson bracket initiated in Chapter 2.

It remains to prove property (6). Fix $\alpha > 0$. Let r^* be the profile function for the pair $(\alpha F, \alpha G)$. It follows from the definition that

$$\alpha r(s) = r^*(\alpha^2 s). \tag{8.1}$$

The triangle inequality for ρ yields

$$|r(t) - r^*(t)| \leq \rho\big((F, G), (\alpha F, \alpha G)\big) = K|1 - \alpha|, \quad K = \|F\| + \|G\|. \tag{8.2}$$

Take any $s, t \in (0, 1)$ and put $\alpha = \sqrt{t/s}$ so that $\alpha^2 s = t$. Combining (8.2) and (8.1), we get that

$$|r(t) - \sqrt{t/s} \cdot r(s)| \leq K|1 - \sqrt{t/s}|.$$

Further, since $r \leq K' := \min(\|F\|, \|G\|)$ we get that

$$|r(t) - r(s)| \leq |r(t) - \sqrt{t/s} \cdot r(s)| + |1 - \sqrt{t/s}| r(s) \leq (K + K')|1 - \sqrt{t/s}|.$$

This yields continuity of r outside 0. $\square$

8.2. Behavior at zero

Recall that by monotonicity of the profile function we can define

$$r^0 := \lim_{s \searrow 0} r(s).$$

The quantity r^0 admits the following upper bound of dynamical flavor. Write f_t for the Hamiltonian flow of F, and put

$$D := \sup_{x,t} |G \circ f_t - G|.$$

The quantity D measures the maximal oscillation of G along the orbits of the Hamiltonian flow of F.

Proposition 8.2.1.

$$r^0 \leq D. \tag{8.3}$$

Proof. For $b > 0$ put

$$G_b := \frac{1}{b} \cdot \int_0^b G \circ f_t \, \mathrm{d}t.$$

Arguing as in the proof of Lemma 7.4.3 we get that

$$|\{F, G_b\}| = \left| \frac{1}{b} \cdot \int_0^b \{F, G \circ f_t\} \, \mathrm{d}t \right| = \left| \frac{1}{b} \cdot \int_0^b \frac{\mathrm{d}}{\mathrm{d}t} G \circ f_t \, \mathrm{d}t \right| = \frac{|G \circ f_b - G|}{b}.$$

This calculation manifests the so-called virial theorem of classical mechanics, see [45]. It follows that $\|\{F, G_b\}\| \leq D/b$. On the other hand,

$$G - G_b = \frac{1}{b} \cdot \int_0^b (G - G \circ f_t) \, \mathrm{d}t,$$

and hence $\|G - G_b\| \leq D$. We conclude that

$$r(D/b) \leq \|F - F\| + \|G - G_b\| \leq D.$$

Passing to the limit as $b \to \infty$ we get $r^0 \leq D$, as required. $\square$

By Property 6, r is continuous on $(0, \infty)$. The behavior of r at 0 is an intriguing problem. Monotonicity of r implies $r(0) \geq r^0$. In dimension two, the converse inequality holds true [34]:

Theorem 8.2.2. *Let M be a closed surface. Then, for all $(F, G) \in \mathcal{F}$, the profile function $r_{F,G}$ is continuous at 0: $r(0) = r^0$.*

Proof. We start with the following observation:

Exercise 8.2.3. Prove that

$$r_{F,G}(\varepsilon) + \sup_{\|\{F_1, G_1\}\| \leq \varepsilon} r_{F_1, G_1}(0) \geq r_{F,G}(0).$$

Recall now Zapolsky's lemma 5.4.26, which in terms of the profile function states that $r_{A,B}(0) \leq c \cdot \sqrt{\|\{A,B\}\|}$ with a universal constant $c = c(M,\omega)$. Below we apply it with $A = F_1$, $B = G_1$. It follows from Exercise 8.2.3 that

$$r_{F,G}(\varepsilon) \geq r_{F,G}(0) - c \cdot \varepsilon^{1/2}, \tag{8.4}$$

and hence r is continuous at 0. □

In higher dimensions, neither a proof of nor a counterexample to the equality $r(0) = r^0$ is currently known.

The following quantitative version of the nondisplaceable fiber theorem (Theorem 6.1.8 above) illustrates the geometric meaning of the value $r(0)$. Consider the map

$$\Phi \colon M \to \mathbb{R}^2, \quad x \mapsto \big(F(x), G(x)\big).$$

Denote by $\mathcal{Q}_a$, $a > 0$ the set of all squares in $\mathbb{R}^2$ whose sides are parallel to the coordinate axes and have length $2a$. Write $\Lambda(F,G)$ for the infimum of $a > 0$ such the preimage $\Phi^{-1}(Q)$ of some square $Q \in \mathcal{Q}_a$ is nondisplaceable.

Proposition 8.2.4.

$$\Lambda(F,G) \leq r(0). \tag{8.5}$$

PROOF. For every $\epsilon > 0$ one can find commuting functions F_0 and G_0 such that

$$\|F - F_0\| + \|G - G_0\| \leq r(0) + \epsilon. \tag{8.6}$$

By Theorem 6.1.8, the map $\Phi_0 = (F_0, G_0)$ from M to $\mathbb{R}^2$ possesses a nonempty nondisplaceable fiber, say $Z := \Phi_0^{-1}(p)$ for some $p \in \mathbb{R}^2$. Consider the square $Q \in \mathcal{Q}_{r(0)+\epsilon}$ with center at p. Observe that $Z \subset \Phi^{-1}(Q)$ in view of inequality (8.6). This readily yields (8.5). □

8.3. A lower bound via symplectic quasi-states

Let (M,ω) be a closed symplectic manifold admitting a subadditive spectral invariant with bounded spectral pseudo-norm. Denote by ζ the corresponding symplectic quasi-state. Recall that in this case the Poisson bracket inequality (4.16) reads

$$\Pi(A,B) := |\zeta(A+B) - \zeta(A) - \zeta(B)| \leq \sqrt{2C \cdot \|\{A,B\}\|} \quad \forall A, B \in C^\infty(M), \tag{8.7}$$

where C is the diameter of $\widetilde{\mathrm{Ham}}(M,\omega)$ with respect to the spectral pseudo-norm.

Proposition 8.3.1 ([66]).

$$r(s) \geq \Pi(F,G)/2 - \sqrt{Cs/2}. \tag{8.8}$$

PROOF. Take any pair of functions A, B with $\|\{A, B\}\| \leq s$ and

$$\|F - A\| + \|G - B\| = \alpha.$$

Since ζ is 1-Lipshitz in the uniform norm,

$$\Pi(A, B) \geq \Pi(F, G) - 2\alpha.$$

Applying (8.7) we get that $\Pi(A, B) < \sqrt{2Cs}$. Combining these inequalities we obtain $\alpha \geq \Pi(F, G)/2 - \sqrt{Cs/2}$, which yields (8.8). □

Example 8.3.2 ([65]). Let $S^2 = \{x^2 + y^2 + z^2 = 1\}$ be the unit sphere equipped with the standard area form. Let ζ be the median quasi-state. Put $F(x, y, z) = 2x^2 - 1$ and $G(x, y, z) = 2y^2 - 1$. Then $F + G = -2z^2$. Calculating (see the discussion after Proposition 5.1.9 above) we get that $\Pi(F, G) = 2$, and hence $r(0) \geq r^0 \geq 1$. At the same time the distance from (F, G) to the commuting pair $(0, G)$ equals 1, and hence $r(0) \leq 1$. We conclude that $r(0) = r^0 = 1$, and therefore the profile function is continuous at 0.

8.4. A lower bound via pb_4

Let (M, ω) be a symplectic manifold and $U \subset M$ be an open subset. Consider a quadruple X_0, X_1, Y_0, Y_1 of compact subsets in U satisfying the intersection condition

$$X_0 \cap X_1 = Y_0 \cap Y_1 = \varnothing.$$

Write $p = \mathrm{pb}_4(X_0, X_1, Y_0, Y_1; U)$ for the pb_4-invariant of the quadruple in U.

Consider any pair of functions $F, G \in C_c^\infty(U)$ from $\mathcal{F}_4(X_0, X_1, Y_0, Y_1)$, that is

$$F|_{X_0} \leq 0, \quad F|_{X_1} \geq 1, \quad G|_{Y_0} \leq 0, \quad G|_{Y_1} \geq 1.$$

Denote by $r = r_{F,G}$ the corresponding profile function. The next result combines ideas from [34] and [33]:

Proposition 8.4.1.

$$r(s) \geq \frac{1}{2} - \frac{s}{2p}. \tag{8.9}$$

PROOF. If $r(s) \geq \frac{1}{2}$ there is nothing to prove. Thus, assume that $r(s) < \frac{1}{2}$. Fix an arbitrary $\alpha \in (r(s), \frac{1}{2})$ and $a > 1$. There exist $A, B \in C_c^\infty(M)$ with

$$\|F - A\| = \alpha_1, \quad \|G - B\| = \alpha_2, \quad \alpha_1 + \alpha_2 \leq \alpha, \quad \|\{A, B\}\| \leq s.$$

Observe that $A \leq \alpha_1$ and $B \leq \alpha_2$ outside U. Furthermore, $A \leq \alpha_1$ on X_0, $A \geq 1 - \alpha_1$ on X_1, $B \leq \alpha_2$ on Y_0, and $B \geq 1 - \alpha_2$ on Y_1.

For $i = 1, 2$, take a nondecreasing function $v_i \colon \mathbb{R} \to [0, 1]$ which vanishes on $(-\infty, \alpha_i)$, equals 1 on $(1 - \alpha_i, +\infty)$, and whose derivative v_i' does not exceed $a/(1-2\alpha_i)$. It follows that the pair of functions $(v_1 \circ A, v_2 \circ B)$ is supported in U and lies in $\mathcal{F}_4(X_0, X_1, Y_0, Y_1)$. Thus

$$\|\{v_1 \circ A, v_2 \circ B\}\| \geq p.$$

On the other hand

$$\begin{aligned}\|\{v_1 \circ A, v_2 \circ B\}\| &= \|(v_1' \circ A) \cdot (v_2' \circ B) \cdot \{A, B\}\| \\ &\leq \frac{a^2 s}{(1-2\alpha_1) \cdot (1-2\alpha_2)} \leq \frac{a^2 s}{1-2\alpha}.\end{aligned}$$

Combining these inequalities we get that

$$\alpha \geq \frac{1}{2} - \frac{a^2 s}{2p}.$$

Since the choice of α and a was arbitrary, we obtain

$$r(s) \geq \frac{1}{2} - \frac{s}{2p},$$

as claimed. □

Example 8.4.2. Assume, in addition, that $0 \leq F \leq 1$. Put $s_0 := \|\{F, G\}\|$ and take any $s \in (0, s_0]$. Take a compactly supported cutoff function $u\colon M \to [0,1]$ which equals 1 on the union of the supports of F and G and which is supported in U. The Poisson bracket of the functions

$$A = u \cdot \big((1 - s/s_0) \cdot 1/2 + s/s_0 \cdot F\big)$$

and $B = G$ satisfies $\|\{A, B\}\| = s$. Therefore,

$$r(s) \leq \|F - A\| + \|G - B\| \leq \frac{1}{2} - \frac{s}{2s_0}.$$

Together with (8.9) this shows that $r(0) = r^0 = \frac{1}{2}$, so that r is continuous at 0.

Example 8.4.3 (Buhovsky's example). Let (M, ω) be any symplectic manifold of dimension ≥ 4. It was shown by Buhovsky [33] that there exists a pair of functions $F, G \in C_c^\infty(M)$ with $r_{F,G}(0) = \frac{1}{2}$ and $\|\{F, G\}\|$ arbitrarily small. Therefore, in contrast to Zapolsky's lemma 5.4.26, on higher-dimensional symplectic manifolds *almost commuting functions are not necessarily nearly commuting.* Below we present Buhovsky's construction in the simplest case when $\dim M = 4$ (the argument in the general case goes through without essential changes). We start with the following exercise.

Exercise 8.4.4. Show that every symplectic manifold of dimension 4 contains an embedded symplectic disc of an arbitrarily large symplectic area. *Hint*: By Darboux's theorem, it suffices to produce such a disc D in the symplectic ball $B \subset \mathbb{R}^4 = \mathbb{C}^2$ of radius $\sqrt{2}$. Fix $N \in \mathbb{N}$ large enough and take D to be the image of the map $z \mapsto (z^N, z)$ of the unit disc $\{|z| < 1\} \subset \mathbb{C}$.

Fix $S > 0$, and let $D \subset M$ be an embedded symplectic disc of symplectic area $> 2S$. By the symplectic neighborhood theorem (see [107, Section 3.3]), a neighborhood of D contains an open subset symplectomorphic to the product of two standard symplectic discs $B^2(r) \times B^2(\delta)$ of radii $r = \sqrt{2S/\pi}$ and $\delta > 0$, equipped with the split symplectic form. The disc $B^2(\delta)$ contains an annulus W_ϵ symplectomorhic to $\{|p| < \epsilon\} \subset T^*S^1$ for every positive $\epsilon < \delta/2$. Denote by $K \subset W_\epsilon$ the simple closed curve corresponding to the zero section of T^*S^1 under this identification. Put $U := B^2(r) \times W_\epsilon$. Take a quadrilateral $\Pi \subset B^2(r)$ of area S with sides a_1, a_2, a_3, a_4, as in Section 7.5.3. Put

$$X_0 = a_1 \times K, \quad X_1 = a_3 \times K, \quad Y_0 = a_2 \times K, \quad Y_1 = a_4 \times K.$$

By (7.18), $\mathrm{pb}_4(X_0, X_1, Y_0, Y_1; U) = 1/S$. Fix now any $\kappa > 1$. By Proposition 7.1.1, there exists a pair of smooth functions $F, G\colon M \to [0,1]$ which are supported in U so that $(F, G) \in \mathcal{F}_4(X_0, X_1, Y_0, Y_1)$ and $\|\{F, G\}\| < \kappa/S$. Therefore, by increasing S we can make $\|\{F, G\}\|$ arbitrarily small. At the same time, by Proposition 8.4.1 and Example 8.4.2, $r_{F,G}(0) = \frac{1}{2}$. This completes Buhovsky's construction.

Let us mention that the question of whether almost commutativity yields near commutativity has been studied for Hermitian matrices. In a fixed dimension n the answer is affirmative [128]: there exists a constant $C_n > 0$ such that for any pair of Hermitian matrices A, B there exist commuting matrices A', B' with

$$\|A - A'\|_{\mathrm{op}} + \|B - B'\|_{\mathrm{op}} \le C_n \sqrt{\|[A, B]\|_{\mathrm{op}}}.$$

It is still unknown whether the constant C_n can be taken to be independent of the dimension n.

CHAPTER 9

Geometry of Covers and Quantum Noise

The nondisplaceable fiber theorem implies that no partition of unity subordinated to a finite cover of a closed symplectic manifold by displaceable open sets can be Poisson commuting (see Corollary 6.1.9 above). In the present chapter we quantify this result and discuss its links to quantum mechanics. In the course of the exposition, we discuss necessary preliminaries on the Berezin–Toeplitz quantization and positive operator valued measures.

9.1. Prelude: covers vs. packings in symplectic topology

Consider B^{2n} an open Euclidean ball in the symplectic vector space $(\mathbb{R}^{2n}, \omega_0)$. Its image under a symplectic embedding into a symplectic manifold (M^{2n}, ω) is called *a standard symplectic ball* in M. An obvious restriction on such a ball comes from the symplectic volume: $\mathrm{vol}(M) \geq \mathrm{vol}(B)$. Gromov's celebrated nonsqueezing theorem provides more subtle constraints. Suppose $n \geq 2$ and consider the Euclidean ball $B^{2n}(R) \subset \mathbb{R}^{2n}$, and the cylinder

$$C^{2n}(r) = B^2(r) \times \mathbb{R}^{2n-2} \subset \mathbb{R}^2(p_1, q_1) \times \mathbb{R}^{2n-2}(p_2, q_2, \ldots, p_n, q_n) = \mathbb{R}^{2n}.$$

The nonsqueezing theorem asserts that if the radii satisfy $R > r$ there is no symplectic embedding $B^{2n}(R) \hookrightarrow C^{2n}(r)$. Note that for any r, R there certainly exists such a volume-preserving embedding. In fact, from the nonsqueezing theorem one can deduce the C^0-rigidity of symplectomorphisms stated in Theorem 1.1.1. Gromov first proved the nonsqueezing theorem using J-holomorphic curves in his 1985 seminal paper [76]. Later on various other proofs were found (see [107, Section 12.1]).

In the same paper Gromov addressed the symplectic packing problem: Find constraints, beyond the volume, on N equal pairwise disjoint standard symplectic balls in a given symplectic manifold. For the sake of further discussion, we present the following sample result (see [23, 87] for surveys on symplectic packings). The symplectic four manifold

$$M_k = \left(S^2 \times S^2, k\omega \oplus \frac{1}{k} \cdot \omega\right), \quad k \in \mathbb{N},$$

where ω is the symplectic form of area 1, contains a standard symplectic ball of volume $1/N_k$ with $N_k = 2k^2 + 2k + 1$. However, *there exists a positive number $v < 1/N_k$ such that M_k does not contain a disjoint union of N_k*

standard symplectic balls of volume v. This result, which manifests rigidity of symplectic packings, follows from Biran's work [22]. Its proof involves Gromov's pseudo-holomorphic curves.

In contrast to packings, covers by open balls exhibit flexible behavior. By using technique due to Rudyak and Schlenk [141] one can prove the following result. *There exists a sequence of positive integers C_n, $n \in \mathbb{N}$ such that for every $N \geq C_n$ and every closed symplectic manifold M^{2n} containing a standard symplectic ball B of volume $v > 1/N$, M admits a cover by N standard symplectic balls of volume v.* The very rough idea is as follows. Cover M by N subsets X_i, $i = 1, \dots, N$ so that each X_i is the union of small pairwise disjoint standard cubes of total volume $< v$. The construction of X_i's involves a careful choice of Darboux charts on M. Next, one finds a Hamiltonian diffeomorphism ϕ_i of M which sends X_i to B. The cover $\{\phi_i^{-1}(B)\}$ does the job.

Interestingly enough, even though constructions of symplectic covers are soft, the covers themselves possess "hard" symplectic invariants which we discuss in this chapter.

9.2. A Poisson bracket invariant of covers

Denote by Q_N the cube $[-1,1]^N \subset \mathbb{R}^N$. For a partition of unity $\vec{F} = \{F_i\}$, $i = 1, \dots, N$, introduce the magnitude of its Poisson noncommutativity

$$\kappa_{\mathrm{cl}}(\vec{F}) := \max_{x \in Q_N, y \in Q_N} \left\| \left\{ \sum_i x_i F_i, \sum_i y_i F_i \right\} \right\|, \tag{9.1}$$

(the subindex cl stands for classical). Given an open cover $\mathcal{U} = \{U_1, \dots, U_N\}$ of M, define the Poisson bracket invariant $\mathrm{pb}(\mathcal{U}) = \inf \kappa_{\mathrm{cl}}(\vec{F})$, where the infimum is taken over all partitions of unity $\vec{F}$ subordinated to $\mathcal{U}$.

Exercise 9.2.1. Show that pb is monotone with respect to refinements: $\mathrm{pb}(\mathcal{V}) \geq \mathrm{pb}(\mathcal{U})$ whenever $\mathcal{V}$ is a refinement of $\mathcal{U}$.

Suppose now that all U_i's are displaceable. Fix a subadditive spectral invariant c on M and put $w(\mathcal{U}) := \max_i w(U_i)$, where $w(U_i)$ stands for the spectral width of U_i associated to c. Displaceability of U_i yields finiteness of $w(\mathcal{U})$ by inequality (4.9). Our next result is a quantitative version of Corollary 6.1.9 above.

Theorem 9.2.2. *Let $\mathcal{U} = \{U_1, \dots, U_N\}$ be a finite open cover of M by displaceable sets. Then*

$$\mathrm{pb}(\mathcal{U}) \cdot w(\mathcal{U}) \geq K(\mathcal{U}), \tag{9.2}$$

where $K(\mathcal{U}) = 1/(2N^2)$.

Proof. The proof is based on the Poisson bracket inequality (4.13) for the partial symplectic quasi-state ζ associated to c. Let $\vec{F}$ be a partition of unity subordinated to $\mathcal{U}$. Put $G_k := \sum_{i=1}^k F_i$. By the vanishing property of ζ combined with the Poisson bracket inequality, it holds that $\zeta(G_1) = 0$ and

$$\zeta(G_{k+1}) - \zeta(G_k) \le \sqrt{2w(\mathcal{U})\kappa_{\mathrm{cl}}(\overrightarrow{F})} \quad \forall k = 1, \ldots, N-1.$$

Since $G_N = 1$, we get that

$$1 = \zeta(G_N) \le N \cdot \sqrt{2w(\mathcal{U})\kappa_{\mathrm{cl}}(\overrightarrow{F})},$$

which yields (9.2). □

A few comments are in order. First, observe that the nonexistence of a Poisson commuting partition of unity subordinated to $\mathcal{U}$ is an immediate consequence of the nondisplaceable fiber theorem (Theorem 6.1.8 above): Indeed, viewing $\overrightarrow{F}$ as a map $M \to \mathbb{R}^N$, we see that each fiber $\overrightarrow{F}^{-1}(p)$ is necessarily contained in some U_i, and hence is displaceable. Thus Theorem 9.2.2 can be consider as a quantitative version of the nondisplaceable fiber theorem in the context of covers. Second, the constant $K(\mathcal{U})$ in the right-hand side of (9.2) seems to be not optimal. Conjecturally, the dependence on N can be eliminated. We refer to [134] for an extensive discussion. Third, Theorem 9.2.2 admits a quantum-mechanical interpretation [133, 134]. The discussion of this interpretation occupies the rest of the chapter.

9.3. The Berezin–Toeplitz quantization

Let (M, ω) be a closed symplectic manifold such that the form $\omega/2\pi$ represents an integral de Rham cohomology class. In what follows such symplectic manifolds will be called *quantizable*. Recall that for a complex Hilbert space H, we denote by $\mathcal{L}(H)$ the space of all bounded Hermitian operators on H. The Berezin–Toeplitz quantization [21, 25, 29, 81, 99, 142] consists of a sequence of finite-dimensional complex Hilbert spaces H_m, $m \to \infty$, of increasing dimension and a family of surjective $\mathbb{R}$-linear maps $T_m \colon C^\infty(M) \to \mathcal{L}(H_m)$ with the following properties:

(BT1) $T_m(1) = \mathbb{1}$;
(BT2) $T_m(F) \ge 0$ for $F \ge 0$;
(BT3) $\|T_m(F)\|_{\mathrm{op}} = \|F\| + O(1/m)$;
(BT4) **(The correspondence principle)**

$$\|-\mathrm{i} \cdot m[T_m(F), T_m(G)] - T_m(\{F, G\})\|_{\mathrm{op}} = O(1/m);$$

(BT5) $\|T_m(F^2) - T_m(F)^2\|_{\mathrm{op}} = O(1/m)$,

as $m \to \infty$ for all $F, G \in C^\infty(M)$. Here $\|A\|_{\mathrm{op}}$ stands for the operator norm of $A \in \mathcal{L}(H)$ and $[A, B]$ for the commutator $AB - BA$. The number m plays the role of the quantum number, while the Planck constant $\hbar$ equals $1/m$, so that $m \to \infty$ is the classical limit. Let us emphasize that the remainders $O(1/m)$ in the formulas above do depend on the norms of the derivatives of the functions F and G. In particular, (BT3)–(BT5) remain valid when

F, G vary smoothly in finite-dimensional families with a compact parameter space.

As an immediate consequence of (BT3) and (BT4) we get that

$$\begin{aligned} \|m[T_m(F), T_m(G)]\|_{\text{op}} &= \|T_m(\{F, G\})\|_{\text{op}} + O(1/m) \\ &= \|\{F, G\}\| + O(1/m) \end{aligned} \tag{9.3}$$

for all smooth functions F, G on M.

Example 9.3.1. Let (M^{2n}, ω, J) be a closed quantizable Kähler manifold. Choose a holomorphic Hermitian line bundle (L, h) over M such that the curvature Θ of its (unique) Hermitian connection compatible with the holomorphic structure equals $-\mathrm{i}\omega$ (existence of (L, h) is a well-known fact from complex algebraic geometry, see, e.g., [158, Section 7.1.3]). The first Chern class of such a bundle is represented by the form $(\mathrm{i}/2\pi)\Theta = (1/2\pi)\omega$. Let V_m be the L^2-space of sections of $L^{\otimes m}$ equipped with the scalar product

$$\langle s_1, s_2 \rangle := \int_M h^{\otimes m}(s_1, s_2)\omega^n.$$

Define $H_m \subset V_m$ as the space of all holomorphic sections of $L^{\otimes m}$. Write Π for the orthogonal projection $V_m \to H_m$. For a function $F \in C^\infty(M)$ denote by $S_F \colon V_m \to V_m$ the multiplication operator $s \mapsto Fs$. With this notation define the operator $T_m(F)$ as $\Pi S_F \Pi^* \colon H_m \to H_m$. It turns out that T_m, $m \to \infty$, is a Berezin–Toeplitz quantization. Properties (BT1) and (BT2) are obvious, while (BT3), (BT4) and (BT5) require a delicate analysis.

The following elementary observation is crucial for our purposes: Let $\overrightarrow{F} = \{F_i\}$, $i = 1, \ldots, N$, be a partition of unity on a closed quantizable symplectic manifold (M, ω). Then the Berezin–Toeplitz quantization takes it to a collection of positive Hermitian operators $\{T_m(F_i)\}$ which sum up to $\mathbb{1}$. Such a collection is an example of a *positive operator valued measure*, abbreviated to POVM, a basic notion of operational quantum mechanics which will be expanded upon in Section 9.4.1. Furthermore, define the magnitude of noncommutativity of this collection of operators by

$$\kappa_{\mathrm{q}}(\{T_m(F_i)\}) := \max_{x,y \in Q_N} \left\| \left[\sum_i x_i T_m(F_i), \sum_i y_i T_m(F_i) \right] \right\|_{\text{op}} \tag{9.4}$$

(compare with (9.1) above, the subindex q stands for quantum). Then by (9.3) we get that if the partition of unity $\overrightarrow{F}$ is subordinated to an open cover $\mathcal{U}$ of M by displaceable sets,

$$\kappa_{\mathrm{q}}(\{T_m(F_i)\}) \geq \frac{\mathrm{pb}(\mathcal{U})}{m} + O(1/m^2) \geq \frac{K}{m} \tag{9.5}$$

for all $m \geq m_0$ for some $K > 0$ which depends only on the cover U. Our objective is to unveil the physical meaning of the magnitude of noncommutativity $\kappa_{\mathrm{q}}(\{T_m(F_i)\})$. As we shall see in Section 9.4.7 below, it is related to the unsharpness principle for POVMs, which can be considered as a cousin of the uncertainty principle in quantum mechanics.

9.4. Operational quantum mechanics

9.4.1. Preliminaries on POVMs. Let H be a complex Hilbert space. Recall that $\mathcal{L}(H)$ denotes the space of all bounded Hermitian operators on H. Consider a set Ω equipped with a σ-algebra $\mathcal{C}$ of its subsets. An $\mathcal{L}(H)$-valued *positive operator valued measure* A on $(\Omega, \mathcal{C})$ is a countably additive map $A\colon \mathcal{C} \to \mathcal{L}(H)$ which takes each subset $X \in \mathcal{C}$ to a positive operator $A(X) \in \mathcal{L}(H)$ and which is normalized by $A(\Omega) = \mathbb{1}$.

POVMs naturally appear in quantum measurement theory [37], where they play the role of generalized observables. The space Ω is called the *value space* of the observable. Pure states of the system are represented by the points of the projective space $[\xi] \in \mathbb{P}(H)$, where $\xi \in H$ is a unit vector. From the viewpoint of von Neumann's axioms of quantum mechanics discussed in Section 5.2 above, a pure state $[\xi]$ corresponds to the orthogonal projector to the line $[\xi] = \mathbb{C} \cdot \xi$. The following is the basic statistical postulate of POVMs:

When the system is in a state $[\xi]$, *the probability of finding the observable* A *in a subset* $X \in \mathcal{C}$ *is postulated to be* $\langle A(X)\xi, \xi\rangle$.

Given an $\mathcal{L}(H)$-valued POVM A on $(\Omega, \mathcal{C})$ and a bounded measurable function $x\colon \Omega \to \mathbb{R}$, one can define the integral $I := \int_\Omega x \, \mathrm{d}A \in \mathcal{L}(H)$. By definition, $\langle I\xi, \xi\rangle = \int_\Omega x \, \mathrm{d}\mu_\xi$, where $\xi \in H$, $|\xi| = 1$ and $\mu_\xi(X) = \langle A(X)\xi, \xi\rangle$, see [20].

Example 9.4.1 (POVMs on finite sets). When $\Omega = \Omega_N := \{1, \dots, N\}$ is a finite set, any POVM A on Ω is fully determined by the N positive Hermitian operators $A_i := A(\{i\})$ which sum up to $\mathbb{1}$.

Example 9.4.2 (Projector valued measures). An important class of POVMs is formed by the *projector valued measures*, for which all the operators $A(X)$, $X \in \mathcal{C}$ are orthogonal projectors. In the language of quantum measurement theory they are called *sharp* observables. Every projector valued measure satisfies $A(X)A(Y) = A(X \cap Y) = A(Y)A(X)$ (we leave this statement as an exercise).

Example 9.4.3 (von Neumann observables). Assume for simplicity that H is finite dimensional. Every "usual" von Neumann observable $A \in \mathcal{L}(H)$ corresponds to the projector valued measure $\hat{A} = \sum P_j \delta_{\lambda_j}$ on $\mathbb{R}$, where $A = \sum_j \lambda_j P_j$ is the spectral decomposition of A. In this case the statistical postulate above agrees with the one of von Neumann's quantum mechanics.

Example 9.4.4 (The Berezin–Toeplitz quantization). The Berezin–Toeplitz quantization can be described in the language of POVMs: There exists a sequence of $\mathcal{L}(H_m)$-valued POVMs B_m on the symplectic manifold M equipped with the Borel σ-algebra so that

$$T_m(f) = \int_M f \, \mathrm{d}B_m. \tag{9.6}$$

For the sake of completeness, let us sketch the argument following the proof of Proposition 1.4.8 of Chapter II in [97]. Indeed, for every vector $\xi \in H_m$ define a linear real-valued functional on $C^\infty(M)$ by $f \mapsto \langle T_m(f)\xi, \xi\rangle$. Since every positive linear functional on $C^\infty(M)$ is given by integration against a Borel measure, there exists a measure $\mu_\xi^{(m)}$ on M with $\langle T_m(f)\xi, \xi\rangle = \int f \, d\mu_\xi^{(m)}$. The POVM B_m is defined by the equality $\langle B_m(X)\xi, \xi\rangle = \mu_\xi^{(m)}(X)$ for every Borel subset $X \subset M$ and vector $\xi \in H_m$. Let us mention also that in the case of Kähler manifolds it is not hard to express the POVMs B_m in terms of coherent states (see [21, 97, 142]).

9.4.2. Naimark's dilation theorem. Suppose that we are given a pair $H \subset H'$ of Hilbert spaces, and an $\mathcal{L}(H')$-valued projector valued measure P on $(\Omega, \mathcal{C})$. Let $\Pi\colon H' \to H$ be the orthogonal projection. Then $A(X) := \Pi P(X) \Pi^*$ is an $\mathcal{L}(H)$-valued POVM on $(\Omega, \mathcal{C})$. We say that P is a *projector valued dilation* of A. A fundamental fact about POVMs, discovered by Naimark (see [116, 153], an alternative spelling is Neumark) is that *every POVM admits a projector valued dilation.*

Example 9.4.5 (Restriction to subspaces). Suppose that P is a projector valued measure on $(\Omega, \mathcal{C})$ with values in $\mathcal{L}(H')$. It is instructive to keep in mind the case when Ω is a finite set, so that P corresponds to a von Neumann observable as in Example 9.4.3 above. We wish to restrict the statistical procedure determined by P to a subspace $H \subset H'$. In other words, given a pure state $\xi \in H$, $|\xi| = 1$, the probability of finding the system in a subset $X \in \mathcal{C}$ equals $p := \langle P(X)\xi, \xi\rangle$. Let us express p in terms of ξ and an operator $A(X) \in \mathcal{L}(H)$. To this end, denote by $\Pi\colon H' \to H$ the orthogonal projector and note that since $\Pi^*\xi = \xi$ for $\xi \in H$,

$$p = \langle P(X)\Pi^*\xi, \Pi^*\xi\rangle = \langle \Pi P(X)\Pi^*\xi, \xi\rangle.$$

Thus, setting

$$A(X) := \Pi P(X) \Pi^* \tag{9.7}$$

we get the desired expression $p := \langle A(X)\xi, \xi\rangle$. Clearly, $A(X)$ is unique, and in general is not a projector. However, $A(X)$, $X \in \mathcal{C}$, is an $\mathcal{L}(H)$-valued POVM. This consideration motivates the appearance of POVMs in quantum mechanics, where restriction of a projector valued measures to a subspace is a natural ingredient of the quantum measurement procedure [37]. Naimark's theorem states that every POVM is the restriction (in the above sense) of some projector valued measure.

Let us slightly reformulate and prove Naimark's theorem.

Theorem 9.4.6 (Naimark dilation theorem). *Let H be a Hilbert space, and let A be an $\mathcal{L}(H)$-valued POVM on $(\Omega, \mathcal{C})$. Then there exist another Hilbert space H', an isometric monomorphism $\phi\colon H \to H'$, and a projector $\mathcal{L}(H')$-valued measure P on $(\Omega, \mathcal{C})$, such that*

$$A(X) = \phi^* P(X) \phi \quad \forall X \in \mathcal{C}. \tag{9.8}$$

Observe that

$$\Pi := \phi\phi^*\colon H' \to H'$$

is the orthogonal projection to the subspace $\phi(H)$, and $\phi^*\phi = \mathbb{1}_H$. Thus, after the identification between H and $\phi(H)$ with the help of ϕ, formula (9.7) agrees with (9.8).

PROOF. (1) Denote by $\mathcal{E}$ the complex vector space consisting of all *simple* functions $f\colon \Omega \to H$, that is, of measurable functions with a finite image. Note that

$$f = \sum_{\xi \in H} \chi_{f^{-1}(\xi)} \cdot \xi, \tag{9.9}$$

where χ_X stands for the indicator function of the subset $X \in \mathcal{C}$.

For a pair of functions $g, h \in \mathcal{E}$ the expression

$$\langle f, g\rangle_\dagger := \sum_{\xi,\eta \in H} \langle A\big(f^{-1}(\xi) \cap g^{-1}(\eta)\big)\xi, \eta\rangle \tag{9.10}$$

defines a nonnegative Hermitian product on $\mathcal{E}$. It degenerates on the subspace $\mathcal{E}_0$ consisting of all functions f with $A\big(f^{-1}(\xi)\big)\xi = 0$ for all $\xi \in H$. Thus it descends to a genuine Hermitian product, which we still denote by $\langle\cdot,\cdot\rangle_\dagger$, on the quotient $\mathcal{E}/\mathcal{E}_0$. Define the Hilbert space H' as the completion of $\mathcal{E}/\mathcal{E}_0$ with respect to the corresponding norm. The map $\phi\colon H \to H'$ which sends a vector $\xi \in H$ to the equivalence class $[c_\xi]$ of the constant function $c_\xi(x) \equiv \xi$ is an isometric monomorphism with respect to $\langle\cdot,\cdot\rangle_\dagger$.

(2) One readily checks that the vector space $\mathcal{E}_0$ is invariant under multiplication by indicator functions χ_X, $X \in \mathcal{C}$. This follows from the description of $\mathcal{E}_0$ given above and the next exercise.

Exercise 9.4.7. Assume that $A(Y)\xi = 0$ for some $Y \in \mathcal{C}$ and $\xi \in H$. Prove that for every subset $Z \subset Y$, $Z \in \mathcal{C}$, one has $A(Z)\xi = 0$. *Hint*: Use the positivity of the POVM A.

Therefore the map

$$P(X)\colon H' \to H', \quad [f] \to [\chi_X \cdot f]$$

is well defined. Note that $P(X)^2 = P(X)$ and

$$\langle P(X)[f], [g]\rangle_\dagger = \langle [f], P(X)[g]\rangle_\dagger = \sum_{\xi,\eta\in H} \langle A\big(X \cap f^{-1}(\xi) \cap g^{-1}(\eta)\big)\xi, \eta\rangle \tag{9.11}$$

for all $[f], [g] \in H'$. Hence, $P(X)$ is an orthogonal projector for every $X \in \mathcal{C}$, and P is a projector $\mathcal{L}(H')$-valued measure on $(\Omega, \mathcal{C})$.

Applying (9.11) we calculate that for all $\xi, \eta \in H$

$$\langle \phi^* P(X)\phi\xi, \eta\rangle = \langle P(X)[c_\xi], [c_\eta]\rangle_\dagger = \langle A(X)\xi, \eta\rangle,$$

which yields (9.8). This completes the proof. $\square$

For future reference let us indicate a number of properties of Naimark's dilation. Here and below End(H) stands for the space of all bounded linear operators on H.

Exercise 9.4.8. In the notations above put

$$\Phi\colon \operatorname{End}(H') \to \operatorname{End}(H), \quad S \mapsto \Pi S \Pi^*. \tag{9.12}$$

Prove that for every bounded measurable function $x\colon \Omega \to \mathbb{R}$

$$\Phi\Big(\int_\Omega x\,\mathrm{d}P\Big) = \int_\Omega x\,\mathrm{d}A, \tag{9.13}$$

where $A(X) = \Phi\big(P(X)\big)$.

Following Janssens [90], define a pairing

$$(\cdot,\cdot)\colon \operatorname{End}(H') \times \operatorname{End}(H') \to \operatorname{End}(H), \quad (S,T) := \Phi(S^*T) - \Phi(S^*)\Phi(T).$$

Exercise 9.4.9. Prove that $(S,T) = g(S)^*g(T)$, where $g(S) = (1-\Pi)S\Pi^*$. Conclude that $(S,S) \geq 0$ for every S.

Theorem 9.4.10.

$$\langle (S,S)\xi,\xi\rangle^{1/2} \cdot \langle (T,T)\xi,\xi\rangle^{1/2} \geq |\langle (S,T)\xi,\xi\rangle| \tag{9.14}$$

for all $S,T \in \operatorname{End}(H')$ *and* $\xi \in H$.

PROOF. By Exercise 9.4.9, for every $\xi \in H$

$$|\langle (S,T)\xi,\xi\rangle| = |\langle g(T)\xi, g(S)\xi\rangle| \leq |g(T)\xi| \cdot |g(S)\xi|.$$

Further,

$$|g(S)\xi| = \langle g(S)^*g(S)\xi,\xi\rangle^{1/2} = \langle (S,S)\xi,\xi\rangle^{1/2},$$

and a similar equality holds for $|g(T)\xi|$. Therefore

$$|\langle (S,T)\xi,\xi\rangle| \leq \langle (S,S)\xi,\xi\rangle^{1/2} \cdot \langle (T,T)\xi,\xi\rangle^{1/2}.$$ □

Corollary 9.4.11. *Let* S,T *be commuting Hermitian operators on* H'. *Then*

$$\langle (S,S)\xi,\xi\rangle^{1/2} \cdot \langle (T,T)\xi,\xi\rangle^{1/2} \geq \tfrac{1}{2} \cdot |\langle [\Phi(S),\Phi(T)]\xi,\xi\rangle|. \tag{9.15}$$

PROOF. Since S and T commute, $[\Phi(S),\Phi(T)] = (T,S) - (S,T)$. The result immediately follows from (9.14). □

9.4.3. The noise operator. Let $A\colon \mathcal{C} \to \mathcal{L}(H)$ be an $\mathcal{L}(H)$-valued POVM on $(\Omega,\mathcal{C})$. For a bounded measurable function $x\colon \Omega \to \mathbb{R}$, put $A(x) := \int x\,\mathrm{d}A$ and define the noise operator by

$$\Delta_A(x) = \int_\Omega x^2\,\mathrm{d}A - A(x)^2$$

(see [38, §2; 39, §3; 102, §2; 126, §4]). Roughly speaking, $A(x)$ is the operator valued expectation of the random variable x with respect to POVM A, while $\Delta_A(x)$ is its operator-valued variance.

Let us give a more precise interpretation of the noise operator (see, e.g., [102]). Take any bounded measurable function x on Ω. Fix a quantum state $\xi \in H$, $|\xi| = 1$. Introduce the following pair of random variables, ϕ and ψ: Consider the measure σ_ξ on Ω given by $\sigma_\xi(X) = \langle A(X)\xi, \xi\rangle$. We define ϕ as $x\colon \Omega \to \mathbb{R}$, where Ω is equipped with the probability measure σ_ξ. The random variable ψ is associated to the von Neumann observable $A(x)$ and the state ξ. It takes the value λ_j with probability $\langle P_j\xi, \xi\rangle$, where $A(x) = \sum \lambda_j P_j$ is the spectral decomposition.

Exercise 9.4.12 ([102])**.** Show that both random variables have the same expectation $\langle A(x)\xi, \xi\rangle$, while the difference of their variances is given in terms of the noise operator:

$$\langle \Delta_A(x)\xi, \xi\rangle = \mathbb{V}(\phi) - \mathbb{V}(\psi).$$

Proposition 9.4.13. *The noise operator is positive*: $\Delta_A(x) \geq 0$.

PROOF. Let an $\mathcal{L}(H')$-projector valued measure P be the Naimark dilation of A (see Section 9.4.2). Recall that this means that H is a subspace in H', $\Pi\colon H' \to H$ is the orthogonal projector, and

$$A(X) = \Phi\big(P(X)\big) \quad \forall X \in \mathcal{C},$$

where

$$\Phi\colon \operatorname{End}(H') \to \operatorname{End}(H), \quad S \mapsto \Pi S \Pi^*.$$

Put $S := \int_\Omega x \, \mathrm{d}P$. Note that $S^2 = \int_\Omega x^2 \, \mathrm{d}P$, since P is projector valued. By (9.13),

$$\Phi(S^2) = \int_\Omega x^2 \, \mathrm{d}A, \quad \Phi(S) = \int_\Omega x \, \mathrm{d}A.$$

Thus

$$\Delta_A(x) = \Phi(S^2) - \Phi(S)^2,$$

and hence is positive by Exercise 9.4.9. □

Exercise 9.4.14. Prove that $\Delta_A(x) = 0$ for all x if and only if A is projector valued.

9.4.4. Uncertainty jump for joint measurements. Let us make a digression which serves as an illustration for various notions introduced earlier Section 9.4. We think of (pure) quantum states as of unit vectors $\xi \in H$. For a von Neumann observable $\hat{A} \in \mathcal{L}(H)$ and a state ξ, write $\mathbb{E}(\hat{A}, \xi) = \langle \hat{A}\xi, \xi\rangle$ for the expectation of $\hat{A}$ in ξ and $\mathbb{V}(\hat{A}, \xi) = \langle \hat{A}^2\xi, \xi\rangle - \mathbb{E}(\hat{A}, \xi)^2$ for the variance. For a pair of observables $\hat{A}, \widehat{B} \in \mathcal{L}(H)$ and a state ξ, the Heisenberg uncertainty principle states that

$$\mathbb{V}(\hat{A}, \xi)\mathbb{V}(\widehat{B}, \xi) \geq \tfrac{1}{4} \cdot |\langle [\hat{A}, \widehat{B}]\xi, \xi\rangle|^2. \tag{9.16}$$

This inequality could be interpreted as follows (see [129, p. 93]): consider an ensemble of quantum particles prepared in the state ξ. For half of the particles measure $\hat{A}$ and for the other half measure $\widehat{B}$. The variances of the corresponding statistical procedures will necessarily satisfy (9.16). Now let us change the setting and try to measure $\hat{A}$ and $\widehat{B}$ simultaneously for *each* of the particles of our ensemble. Even though one cannot do this *precisely* unless $\hat{A}$ and $\widehat{B}$ commute (see [37, p. 28]), an approximate joint measurement does exists. The latter can be described in the language of operational quantum physics as follows.

An *approximate measurement* of a von Neumann observable $\hat{A}$ is given by an $\mathcal{L}(H)$-valued POVM $F = \{F_1, \dots, F_N\}$ on a finite set[1] $\Omega_N = \{1, \dots, N\}$ together with a random variable $x\colon \Omega \to \mathbb{R}$. Let us emphasize that so far $\hat{A}$ and F are unrelated: we made a "declaration of intention." A meaningful link between $\hat{A}$ and F appears in the next definition. We say that the measurement is *unbiased* if the expected value of (F, x) at any state ξ coincides with that of $\hat{A}$, that is $F(x) := \sum x_i F_i = \hat{A}$ (see [38, p. 263]). The expectation of the random variable given by (F, x) and ξ is $\mathbb{E}(F, x, \xi) = \langle F(x)\xi, \xi\rangle$, while the variance has the form

$$\mathbb{V}(F, x, \xi) = \sum x_i^2 \langle F_i \xi, \xi\rangle - \mathbb{E}(F, x, \xi)^2.$$

According to Exercise 9.4.12, the latter variance can be rewritten via the noise operator as

$$\mathbb{V}(F, x, \xi) = \mathbb{V}(\hat{A}, \xi) + \langle \Delta_F(x)\xi, \xi\rangle. \tag{9.17}$$

Since the noise operator $\Delta_F(x)$ is positive, this gives

$$\mathbb{V}(F, x, \xi) \geq \mathbb{V}(\hat{A}, \xi). \tag{9.18}$$

A *joint approximate measurement* of a pair of observables $\hat{A}, \widehat{B} \in \mathcal{L}(H)$ is given by a POVM F together with a pair of random variables x, y on Ω_N such that (F, x) and (F, y) intend to measure $\hat{A}$ and $\widehat{B}$ respectively. Such a measurement is called unbiased if $F(x) = \hat{A}$ and $F(y) = \widehat{B}$. By (9.18) and the Heisenberg principle (9.16), we have

$$\mathbb{V}(F, x, \xi)\mathbb{V}(F, y, \xi) \geq \tfrac{1}{4} \cdot |\langle [\hat{A}, \widehat{B}]\xi, \xi\rangle|^2.$$

Interestingly enough, this uncertainty bound can be improved.

Theorem 9.4.15.

$$\mathbb{V}(F, x, \xi)\mathbb{V}(F, y, \xi) \geq |\langle [\hat{A}, \widehat{B}]\xi, \xi\rangle|^2 \tag{9.19}$$

for every joint unbiased approximate measurement of observables $\hat{A}$ and $\widehat{B}$ prepared in any state ξ.

Every pair of observables $\hat{A}, \widehat{B}$ admits a joint unbiased approximate measurement. Indeed, consider the spectral decompositions $\hat{A} = \sum_{i=1}^{k} a_i P_i$ and $\widehat{B} = \sum_{i=1}^{m} b_i Q_i$. Define the POVM

$$F = \{P_1/2, \dots, P_k/2, Q_1/2, \dots, Q_m/2\} \tag{9.20}$$

on $\Omega = \{1, \dots, k+m\}$. Then F together with the random variables

$$x = (2a_1, \dots, 2a_k, 0, \dots, 0) \quad \text{and} \quad y = (0, \dots, 0, 2b_1, \dots, 2b_m)$$

[1] All the results below remain valid for POVMs on arbitrary spaces.

provides an unbiased approximate joint measurement of $\hat{A}$ and $\widehat{B}$.

Let us focus now on the example where $H = \mathbb{C}^2$ and $\hat{A}, \widehat{B}$ are the Pauli matrices,

$$\hat{A} = \begin{pmatrix} 0 & 1 \\ 1 & 0 \end{pmatrix}, \quad \widehat{B} = \begin{pmatrix} 1 & 0 \\ 0 & -1 \end{pmatrix}.$$

Then for $\xi = 1/\sqrt{2} \cdot (1, \mathrm{i})$ and F as in (9.20), one readily checks that $\mathbb{V}(F, x, \xi) = \mathbb{V}(F, y, \xi) = 2$, $\mathbb{V}(\hat{A}, \xi) = \mathbb{V}(\widehat{B}, \xi) = 1$, and $|\langle[\hat{A}, \widehat{B}]\xi, \xi\rangle|^2 = 4$. Thus in this case both inequalities (9.16) and (9.19) are tight. Theorem 9.4.15 shows that the Heisenberg uncertainty quantified as the product of variances at the state ξ jumps from 1 to ≥ 4 for every unbiased approximate joint measurement of $\hat{A}$ and $\widehat{B}$. This is an example of *the uncertainty jump* appearing in the title of this subsection.

The jump phenomenon had been observed for joint position-momentum measurements by Arthurs and Kelly [10], Stulpe, Gudder, and Hagler [151] (see also the monograph [37] by Busch, Grabowski, and Lahti, p. 63), and in the general case by Arthurs and Goodman [9] (see also Section 3.3 in Ozawa's paper [127]) and by Ishikawa [89, Theorem 3]. We refer to [40] for further discussion.

The proof of Theorem 9.4.15 is based on the following inequality, which can be found in the book [83, Theorem 7.5].

Theorem 9.4.16.

$$\langle\Delta_F(x)\xi, \xi\rangle \cdot \langle\Delta_F(y)\xi, \xi\rangle \geq \tfrac{1}{4} \cdot |\langle[\hat{A}, \widehat{B}]\xi, \xi\rangle|^2. \tag{9.21}$$

Proof. We follow an approach by Janssens [90]. By the Naimark dilation theorem, one can isometrically embed H into a Hilbert space H' and find an $\mathcal{L}(H')$-valued projector valued measure $P = \{P_1, \ldots, P_N\}$ on Ω_N so that $F_i = \Phi(P_i)$, where

$$\Phi\colon \operatorname{End}(H') \to \operatorname{End}(H), \quad X \mapsto \Pi X \Pi^*,$$

where $\Pi\colon H \to H'$ stands for the orthogonal projection. Recall the notation

$$(S, T) := \Phi(ST) - \Phi(S)\Phi(T)$$

for $S, T \in \mathcal{L}(H')$. Put $S = \sum x_i P_i$ and $T = \sum y_i P_i$. Note that $\Phi(S) = \hat{A}$, $\Phi(T) = \widehat{B}$, $(S, S) = \Delta_F(x)$ and $(T, T) = \Delta_F(y)$. The desired inequality immediately follows from Corollary 9.4.11. □

Proof of Theorem 9.4.15. We use that $a + b \geq 2\sqrt{ab}$. By (9.17),

$$\mathbb{V}(F, x, \xi)\mathbb{V}(F, y, \xi) \geq 4 \cdot \sqrt{\mathbb{V}(\hat{A}, \xi) \cdot \langle\Delta_F(x)\xi, \xi\rangle \cdot \mathbb{V}(\widehat{B}, \xi) \cdot \langle\Delta_F(y)\xi, \xi\rangle}.$$

Applying (9.16) and Theorem 9.4.16 we get that

$$\mathbb{V}(F, x, \xi)\mathbb{V}(F, y, \xi) \geq 4 \cdot \sqrt{(1/4)^2 \cdot |\langle[\hat{A}, \widehat{B}]\xi, \xi\rangle|^2} = |\langle[\hat{A}, \widehat{B}]\xi, \xi\rangle|^2,$$

as required. □

Another application of inequality (9.21) is as follows:

Exercise 9.4.17. Let (M,ω) be a quantizable symplectic manifold. Fix a scheme $T_m\colon C^\infty(M) \to \mathcal{L}(H_m)$ of the Berezin–Toeplitz quantization. Prove that for every pair of functions $u, v \in C^\infty(M)$

$$\begin{aligned}(9.22)\quad & \|T_m(u^2) - T_m(u)^2\|_{\mathrm{op}}^{1/2} \cdot \|T_m(v^2) - T_m(v)^2\|_{\mathrm{op}}^{1/2} \\ & \geq \tfrac{1}{2} \cdot \|[T_m(u), T_m(v)]\|_{\mathrm{op}} = \|\{u,v\}\|/(2m) + O(1/m^2).\end{aligned}$$

Conclude that for every nonconstant function u, $T_m(u^2) \neq T_m(u)^2$ for all sufficiently large m. *Hint*: Show that $\Delta_{B_m}(u) = T_m(u^2) - T_m(u)^2$, where B_m is the POVM associated to the Berezin–Toeplitz quantization as in Example 9.4.4 above. Then apply (9.21) and (9.3).

9.4.5. Smearing of POVMs. Let Ω and Θ be Hausdorff, locally compact, second countable topological spaces and let $\mathcal{C}, \mathcal{D}$ be their Borel σ-algebras, respectively. Denote by $\mathcal{P}(\Omega)$ the set of Borel probability measures on Ω. A *Markov kernel* is a map

$$\gamma\colon \Theta \to \mathcal{P}(\Omega), \qquad w \mapsto \gamma_w,$$

such that the function $w \mapsto \gamma_w(X)$ on Θ is measurable for every $X \in \mathcal{C}$. Let A and B be POVMs on $(\Omega, \mathcal{C})$ and $(\Theta, \mathcal{D})$, respectively. We say that A is a *smearing*[2] of B [3, 37, 91] if there exists a Markov kernel γ such that

$$A(X) = \int_\Theta \gamma_w(X)\, \mathrm{d}B(w) \quad \forall X \in \mathcal{C}.$$

In the physical language, each element w of the value set Θ of B diffuses into a subset $X \in \mathcal{C}$ with probability $\gamma_w(X)$.

Denote by $Q(\Omega)$ the set of all measurable functions $x\colon \Omega \to \mathbb{R}$ with $\max|x| \leq 1$ which are interpreted as random variables on Ω. Every Markov kernel γ as above defines a *smearing operator*

$$(9.23)\qquad \Gamma\colon Q(\Omega) \to Q(\Theta), \qquad (\Gamma x)(w) = \int_\Omega x\, \mathrm{d}\gamma_w \quad \forall w \in \Theta.$$

In the physical slang, the random variable Γx is a coarse-graining of x: its value at a point $w \in \Theta$ is the expectation of x with respect to the measure γ_w.

Example 9.4.18 (Measurable partitions of unity). If $\Omega = \Omega_N = \{1, \ldots, N\}$, a Markov kernel γ is given by a collection of nonnegative measurable functions γ_j on Θ, $j = 1, \ldots, N$, so that $\sum_j \gamma_j(w) = 1$ for all $w \in \Theta$, that is, by a measurable partition of unity. A POVM $A = \{A_1, \ldots, A_N\}$ on Ω_N,

$$(9.24)\qquad A_j = \int_\Theta \gamma_j\, \mathrm{d}B,$$

is a smearing of a POVM B on $(\Theta, \mathcal{D})$. Every function x from $Q(\Omega_N)$ is canonically identified with a vector $x = \big(x(1), \ldots, x(N)\big)$ lying in the cube $Q_N := [-1,1]^N$. The smearing operator Γ is given by $\Gamma(x) = \sum_j x_j \gamma_j$.

[2]Some authors call it randomization or fuzzification.

9.4.6. Measuring noise. Now we are ready to introduce one of the central characters of our story.

Definition 9.4.19. The *inherent noise of a POVM A on $(\Omega, \mathcal{C})$ at a pure state $\xi \in H$* is given by

$$\mathcal{N}(A, \xi) := \inf_{B,\Gamma} \sup_{x \in Q(\Omega)} \langle \Delta_B(\Gamma x)\xi, \xi \rangle,$$

where the infimum is taken over all POVMs B such that A is a smearing of B and Γ is the corresponding smearing operator.

Exercise 9.4.20. Prove that

$$0 \le \mathcal{N}(A, \xi) \le 1. \tag{9.25}$$

Proposition 9.4.21 ([101, Lemma 2, p. 277]).

$$\Delta_B(\Gamma x) \le \Delta_A(x), \tag{9.26}$$

where A is a smearing of B with smearing operator Γ and x is any function from $Q(\Omega)$.

Proof. Observe that $B(\Gamma x) = A(x)$. Furthermore, in the notation of Section 9.4.5, one calculates that

$$\int_\Omega x^2 \, \mathrm{d}A - \int_\Theta (\Gamma x)^2 \, \mathrm{d}B = \int_\Theta h(w) \, \mathrm{d}B(w),$$

with

$$h(w) = \int_\Omega x^2 \, \mathrm{d}\gamma_w - \left(\int_\Omega x \, \mathrm{d}\gamma_w \right)^2.$$

Since $h(w) \ge 0$ by the Cauchy-Schwarz inequality, we get (9.26). □

Since smearing can be interpreted as a diffusion, the increment

$$\langle \Delta_A(x)\xi, \xi \rangle - \langle \Delta_B(\Gamma x)\xi, \xi \rangle$$

(which is always nonnegative by (9.26)) plays the role of a random component of the noise of A. This is why we call $\mathcal{N}(A, \xi)$ the *inherent* (as opposed to random) noise of A at the state ξ.

Let us mention that the inherent noise behaves monotonically under smearings.

Proposition 9.4.22. *For every pure state $\xi \in H$, $\mathcal{N}(A, \xi) \le \mathcal{N}(B, \xi)$ whenever A is a smearing of B.*

Proof. Introduce the following notation: For a POVM A we write Ω^A for the value space of A and we abbreviate $Q^A := Q(\Omega^A)$. Following [101], we write $B \to A$ if A is a smearing of B, and denote by $\Gamma_{BA} \colon Q^A \to Q^B$ the smearing operator.

Observe that if $C \to B \to A$ we have that $\Gamma_{CA} = \Gamma_{CB}\Gamma_{BA}$. Therefore,

$$\sup_{x \in Q^A} \langle \Delta_C(\Gamma_{CA} x)\xi, \xi \rangle = \sup_{x \in Q^A} \langle \Delta_C(\Gamma_{CB}\Gamma_{BA} x)\xi, \xi \rangle \le \sup_{y \in Q^B} \langle \Delta_C(\Gamma_{CB} y)\xi, \xi \rangle,$$

and hence

$$\inf_{C:C\to B\to A} \sup_{x\in Q^A} \langle \Delta_C(\Gamma_{CA}x)\xi,\xi\rangle \leq \inf_{C:C\to B} \sup_{y\in Q^B} \langle \Delta_C(\Gamma_{CB}y)\xi,\xi\rangle \leq \mathcal{N}(B,\xi). \tag{9.27}$$

Note that

$$\{C : C \to B \to A\} \subset \{C : C \to A\}.$$

Thus the left hand side of (9.27) is $\geq \mathcal{N}(A,\xi)$. We conclude that $\mathcal{N}(A,\xi) \leq \mathcal{N}(B,\xi)$, as required. □

Finally, define the "maximal" inherent noise of a POVM A as

$$\mathcal{N}(A) = \sup_{\xi} \mathcal{N}(A,\xi), \tag{9.28}$$

where the supremum is taken over all pure states $\xi \in H$, $|\xi| = 1$. Note that this quantity is in general smaller than the inherent noise introduced in [134], and hence the lower bound (9.31) below is, strictly speaking, stronger than the one in [134].

9.4.7. An unsharpness principle. For an $\mathcal{L}(H)$-valued POVM A on Ω and a pure state $\xi \in H$, introduce *the magnitudes of noncommutativity*

$$\kappa_{\mathrm{q}}(A,\xi) = \sup_{x,y\in Q(\Omega)} |\langle [A(x),A(y)]\xi,\xi\rangle|$$

and

$$\kappa_{\mathrm{q}}(A) = \sup_{|\xi|=1} \kappa_{\mathrm{q}}(A,\xi) = \sup_{x,y\in Q(\Omega)} \|[A(x),A(y)]\|_{\mathrm{op}};$$

compare with the definition given in (9.4) above.

Theorem 9.4.23. *For every pure state $\xi \in H$*

$$\mathcal{N}(A,\xi) \geq \tfrac{1}{2}\kappa_{\mathrm{q}}(A,\xi). \tag{9.29}$$

PROOF. Suppose that an $\mathcal{L}(H)$-valued POVM A on $(\Omega,\mathcal{C})$ is a smearing of B with smearing operator Γ. Take any $x,y \in Q(\Omega)$. Since $B(\Gamma x) = A(x)$ and $B(\Gamma y) = A(y)$, from Theorem 9.4.16 we get that

$$\langle \Delta_B(\Gamma x)\xi,\xi\rangle \cdot \langle \Delta_B(\Gamma y)\xi,\xi\rangle \geq \tfrac{1}{4} \cdot |\langle [A(x),A(y)]\xi,\xi\rangle|^2. \tag{9.30}$$

Therefore

$$\sup_{x\in Q(\Omega)} \langle \Delta_B(\Gamma x)\xi,\xi\rangle \geq \tfrac{1}{2}\kappa_{\mathrm{q}}(A,\xi),$$

which yields (9.29). □

As an immediate consequence of Theorem 9.4.23 we get the following result:

Theorem 9.4.24 (Unsharpness principle).

$$\mathcal{N}(A) \geq \tfrac{1}{2}\kappa_{\mathrm{q}}(A). \tag{9.31}$$

It is known [3, Section 5] (cf. [91]) that every commutative POVM on a Hausdorff locally compact second countable space is necessarily a smearing of a sharp observable, that is, of a projector valued measure. In particular, $\mathcal{N}(A) = 0$ if $\kappa_{\mathrm{q}}(A) = 0$. The unsharpness principle above shows that the converse statement is also true.

Exercise 9.4.25. Show that the magnitude of noncommutativity behaves monotonically with respect to smearings (see [133]):

$$\kappa_{\mathrm{q}}(B) \geq \kappa_{\mathrm{q}}(A) \quad \text{if } A \text{ is a smearing of } B. \tag{9.32}$$

Thus, after a smearing both sides of inequality (9.31) decrease.

9.5. Classical and quantum registration procedures

Take a finite cover $\mathcal{U} = \{U_1, \ldots, U_N\}$ of a set M. Consider the following procedure: Each point $z \in M$ has to register in exactly one set U_j containing it. In other words, one has to assign to each $z \in M$ an index $j \in \Omega_N := \{1, \ldots, N\}$ so that $z \in U_j$. This task is in general ambiguous because of the overlaps between the subsets of the cover. In order to resolve the ambiguity, the assignment is made at random: one chooses an index $j \in \Omega_N$ corresponding to a point $z \in M$ with probability $F_j(z)$, where

$$F_j(z) = 0 \quad \text{provided } z \notin U_j. \tag{9.33}$$

The latter condition guarantees a correct answer to the question '*Where (i.e., in which set U_j) is a given point z located?*' Each specific outcome of the registration provides "the truth, but not the whole truth."

As an illustration,[3] consider the following toy model of a cellular communication network consisting of a collection of access points $u_1, \ldots, u_N$. Each access point u_j can be reached from a domain U_j, the so called *location area.* The location areas U_j cover some territory M considered, for simplicity, as a plane domain. Each cell phone at a given location $z \in M$ must register in exactly one access point u_j whose location area U_j contains z. When z lies on the overlap between several location areas, the choice between them is made at random with a probability depending on z (in reality, of course, the choice involves more delicate considerations, see [80]).

Suppose now that M is a topological space, the subsets U_j are open, and the probabilities $F_j(z)$ are continuous functions. Let us slightly strengthen condition (9.33) above by assuming that F_j vanishes in *a neighborhood* of the closed set $M \setminus U_j$. This means that F_j is supported in U_j. Since the probabilities $F_j(z)$, $j = 1, \ldots, N$, are nonnegative and sum up to 1 for every z, the functions F_j form a partition of unity subordinated to the cover $\{U_j\}$. Given a probability measure σ on M, a randomly chosen (with respect to σ) point of M is registered in the set U_j with probability

$$\int_M F_j(z)\, \mathrm{d}\sigma. \tag{9.34}$$

[3]Thanks to Iosif Polterovich for this idea.

Next, let us discuss this registration procedure in the context of *the Hilbert space model of classical mechanics* [111, p. 1628]. Let M be a compact classical phase space whose phase volume is given by a probability measure μ. Consider the Hilbert space $H = L_2(M, \mu)$ of square-integrable complex valued functions on M. The states of the system are represented by unit vectors $\xi \in H$. Each such vector gives rise to a probability measure $\mathrm{d}\sigma_\xi = |\xi|^2 \cdot \mathrm{d}\mu$ on M. Take any partition of unity $\overrightarrow{F} = \{F_1, \dots, F_N\}$ subordinated to an open cover $\mathcal{U} = \{U_1, \dots, U_N\}$ of M and consider the registration procedure associated to this data. According to formula (9.34), if the system is prepared in the state ξ, it is registered in a domain U_j with probability

$$\int_M F_j(z)\, \mathrm{d}\sigma_\xi. \tag{9.35}$$

This statistical procedure can be described by an $\mathcal{L}(H)$-valued POVM $A = \{A_1, \dots, A_N\}$ on Ω_N, where each A_j is the operator of multiplication by F_j. Treating A as an observable, we see that at a state $\xi \in H$ it accepts the value $j \in \Omega_N$ with the probability

$$\langle A_j \xi, \xi \rangle = \int_M F_j |\xi|^2 \, \mathrm{d}\mu = \int_M F_j \, \mathrm{d}\sigma_\xi,$$

which agrees with rule (9.35) above.

Clearly, the POVM A is commutative, and hence it is a smearing of a sharp observable. The latter can be chosen as the following canonical $\mathcal{L}(H)$-valued projector valued measure P on the phase space M. Given a Borel subset $X \subset M$, the projector $P(X)$ is the operator of multiplication by the indicator function χ_X of X. The relation

$$A_j = \int_M F_j(z)\, \mathrm{d}P(z)$$

shows that A is the smearing of P with the Markov kernel given by the partition of unity $\overrightarrow{F}$ (see (9.24) above). Thus the inherent noise $\mathcal{N}(A)$ vanishes.

Finally, let us describe the quantum version of our registration procedure. Let (M^{2n}, ω) be a closed symplectic manifold equipped with the phase volume ω^n, and let $T_m \colon C^\infty(M) \to \mathcal{L}(H_m)$ be a Berezin–Toeplitz quantization. Consider the registration procedure associated to an open cover $\mathcal{U} = \{U_1, \dots, U_N\}$ of M and a subordinated partition of unity $\overrightarrow{F} = \{F_1, \dots, F_N\}$. The *quantum registration* is described by the $\mathcal{L}(H_m)$-valued POVM $A^{(m)}$ formed by the Toeplitz operators $T_m(F_j)$, $j = 1, \dots, N$. In a pure state $\xi \in H_m$, $|\xi| = 1$, the quantum system is registered in a subset U_j with probability $\langle T_m(F_j)\xi, \xi\rangle$. Recall that $\hbar = 1/m$.

Theorem 9.5.1. (i) *There exists a constant $K_+ > 0$ such that the inherent noise of the quantum registration procedure satisfies*

$$K_+ \hbar \geq \mathcal{N}(A^{(m)}) \tag{9.36}$$

for all sufficiently large m. *Here* K_+ *depends on the partition of unity* $\vec{F}$.

(ii) *Suppose that all sets of the cover* $\mathcal{U}$ *are displaceable. Then* $p := \mathrm{pb}(\mathcal{U}) > 0$ *and for every* $K_- < p/2$

$$\mathcal{N}(A^{(m)}) \geq K_-\hbar \tag{9.37}$$

for all sufficiently large m.

PROOF. Observe that the POVM $A^{(m)}$ is the smearing of the POVM B_m associated to the Berezin–Toeplitz quantization, and the smearing operator is given by that partition of unity $\vec{F}$ (see Examples 9.4.4 and 9.4.18 above). Therefore by definition of the inherent noise

$$\mathcal{N}(A^{(m)}) \leq \sup_{x \in Q_N} \left\| \Delta_{B_m}\left(\sum_i x_i F_i\right) \right\|_{\mathrm{op}}.$$

But, abbreviating $F_x := \sum_i x_i F_i$, we have that

$$\Delta_{B_m}(F_x) = T_m(F_x^2) - T_m(F_x)^2 = O(1/m)$$

by property (BT5) of the Berezin–Toeplitz quantization. This proves the upper bound in (i).

The lower bound in (ii) follows from inequality (9.5) combined with the unsharpness principle given by Theorem 9.4.24. □

The registration procedure presented above can be considered as an attempt to localize the mechanical system in the phase space. We conclude that *any phase space localization of the quantized system beyond a certain scale yields an inherent noise.* Interestingly enough, this scale is governed by the notion of displaceability provided by symplectic topology.

9.6. Geometry of overlaps and pb_4

We conclude this chapter with yet another lower bound on $\mathrm{pb}(\mathcal{U})$ coming from quite a subtle geometry of overlaps between the sets of the cover $\mathcal{U}$. The bound is given in terms of the Poisson bracket invariant pb_4 introduced in Section 7.1.

Given a finite open cover $\mathcal{U} = \{U_1, \ldots, U_N\}$ of M, consider any decomposition of the set $\Omega_N = \{1, \ldots, N\}$ into disjoint union $I \sqcup I^{\mathrm{c}}$ (here and below the upper index c stands for the complement). Denote

$$U(I) := \left(\bigcup_{\alpha \in I} U_\alpha\right)^{\mathrm{c}}.$$

We illustrate this in Figure 9.1: Here the cover consists of discs. The discs U_α, $\alpha \in I$, form the left column and the discs U_β, $\beta \in I^{\mathrm{c}}$, form the right column. The sets $U(I)$ and $U(I^{\mathrm{c}})$ appear in gray (on the right) and in white (on the left) respectively.

Theorem 9.6.1. *For every* $I, J \subset \Omega_N$,

$$\mathrm{pb}(\mathcal{U}) \geq 4\,\mathrm{pb}_4\big(U(I), U(I^{\mathrm{c}}), U(J), U(J^{\mathrm{c}})\big). \tag{9.38}$$

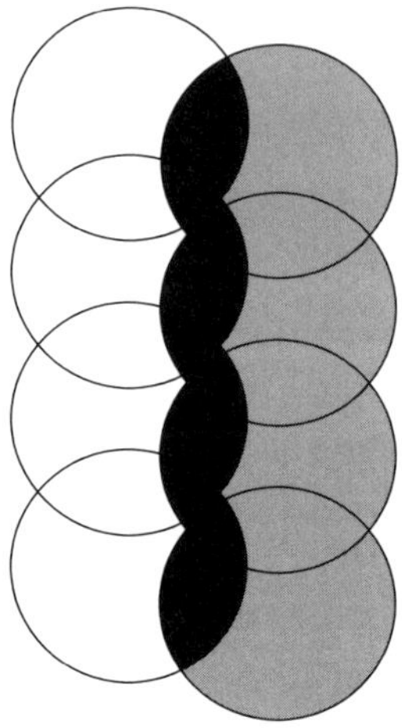

FIGURE 9.1. Geometry of overlaps.

PROOF. Take an arbitrary partition of unity $\overrightarrow{F} = \{F_1, \ldots, F_N\}$ subordinated to $\mathcal{U}$. Put $\phi := \sum_{\alpha \in I} F_\alpha$ and $\psi := \sum_{\alpha \in J} F_\alpha$. Then $1 - \phi = \sum_{\beta \in I^c} F_\beta$ and $1 - \psi = \sum_{\beta \in J^c} F_\beta$. Observe that $\phi = 0$ near $U(I)$, $1 - \phi = 0$ near $U(I^c)$, $\psi = 0$ near $U(J)$ and $1 - \psi = 0$ near $U(J^c)$. By definitions of κ_{cl} and pb_4,

$$\kappa_{\mathrm{cl}}(\overrightarrow{F}) \geq \|\{\phi - (1-\phi), \psi - (1-\psi)\}\| = 4\|\{\phi, \psi\}\| \\ \geq 4\,\mathrm{pb}_4\big(U(I), U(I^c), U(J), U(J^c)\big).$$

Since this holds for every $\overrightarrow{F}$, we get (9.38). □

Example 9.6.2. Consider the unit sphere $S^2 = \{x^2 + y^2 + z^2 = 1\}$ equipped with some area form ω. Assume that the total area of the sphere equals 1. Consider a cover $\mathcal{U}$ of the sphere by four discs

$$U_1 = \{x < \tfrac{1}{2}, y > -\tfrac{1}{2}\}, \qquad U_2 = \{x > -\tfrac{1}{2}, y > -\tfrac{1}{2}\},$$
$$U_3 = \{x > -\tfrac{1}{2}, y < \tfrac{1}{2}\}, \qquad U_4 = \{x < \tfrac{1}{2}, y < \tfrac{1}{2}\}.$$

The intersection of all these discs consists of two quadrilaterals. Denote by Π the one lying in upper hemisphere, that is $\Pi = \{|x| < \frac{1}{2}, |y| < \frac{1}{2}, z > 0\}$ (the projections of these sets onto the $x - y$ plane are illustrated in Figure 9.2). Our objective is to estimate $\mathrm{pb}(U)$ by using inequality (9.38). Taking $I = \{1, 2\}$ and $J = \{1, 4\}$ we get that

$$U(I) = \{y \leq -\tfrac{1}{2}\}, \qquad U(I^c) = \{y \geq \tfrac{1}{2}\},$$
$$U(J) = \{x \geq \tfrac{1}{2}\}, \qquad U(J^c) = \{x \leq -\tfrac{1}{2}\}.$$

Denote by a_0, a_1, b_0, b_1 the sides of Π lying in $U(I), U(I^c), U(J), U(J^c)$, respectively. By monotonicity of pb_4 (see (7.7) above) and Theorem 7.5.3, we get that

$$\mathrm{pb}_4\big(U(I), U(I^c), U(J), U(J^c)\big) \geq \mathrm{pb}_4(a_0, a_1, b_0, b_1) \geq 1/A,$$

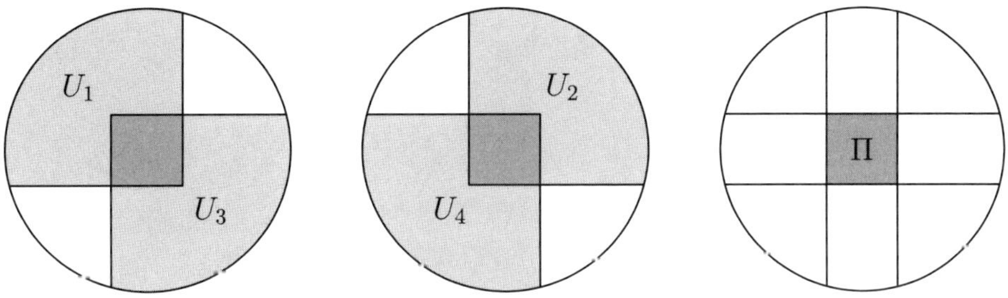

FIGURE 9.2. The cover U_1, U_2, U_3, U_4 of S^2.

where $A = \mathrm{Area}(\Pi)$. Thus, by (9.38),

$$\mathrm{pb}(\mathcal{U}) \geq 4/A. \tag{9.39}$$

Suppose now that the form ω is chosen in such a way that disc U_1 has area $> \frac{1}{2}$ and hence is not displaceable. Then Theorem 9.2.2 involving the magnitude of localization is not applicable, while inequality (9.39) provides a lower bound on pb_4 in terms of geometry of overlaps.

CHAPTER 10

Preliminaries from Morse Theory

The rest of the book is dedicated to the construction of subadditive spectral invariants of Hamiltonian diffeomorphisms, which play a key role in the theory of (partial) symplectic quasi-states. The construction is based on Hamiltonian Floer theory, which will be discussed in the next chapter. To make this theory more accessible, we begin with its classical counterpart, the Morse–Novikov homology of a closed 1-form on a finite-dimensional manifold.

10.1. Spectral numbers of functions

If we were asked to present the construction of subadditive spectral invariants "while standing on one foot," the answer would be that subadditive spectral invariants are the *spectral numbers of the classical action functional on the free loop space associated to a symplectic manifold.* In this section we discuss a toy case of spectral numbers of smooth functions on finite-dimensional manifolds.

Let M^n be a closed n-dimensional manifold. Throughout this section we write H_* for the homology with $\mathbb{Z}_2$-coefficients. For a smooth function $F \in C^\infty(M)$ let

$$i_\alpha \colon H_*(\{F < \alpha\}) \to H_*(M)$$

be the homomorphism induced by the inclusion

$$\{F < \alpha\} \hookrightarrow M.$$

By convention, we set $H(\varnothing) = 0$. Let $a \in H_*(M) \setminus \{0\}$. Define the *spectral number*[1]

$$c(a, F) = \inf\{\alpha : a \in \operatorname{Image}(i_\alpha)\}.$$

Proposition 10.1.1. *The spectral number $c(a, F)$ is a critical value of F.*

Proof. Denote $c = c(a, F)$ and assume on the contrary that c is not a critical value of F. By compactness, we can find $\varepsilon > 0$ such that the interval $[c - \varepsilon, c + \varepsilon]$ contains no critical values of F. But then the inclusion

$$\{F < c - \varepsilon\} \hookrightarrow \{F < c + \varepsilon\}$$

[1] In the terminology of [132], spectral numbers are the homologically essential critical values of F.

is a homotopy equivalence (see [109]). In particular,

$$\mathrm{Image}(i_{c-\varepsilon}) = \mathrm{Image}(i_{c+\varepsilon}).$$

This contradicts the fact that c is a spectral number. Therefore, c must be a critical value. □

Proposition 10.1.2. *The maximum and minimum values of F can be characterized as follows:*

$$\max F = c([M], F), \quad \min F = c([\mathrm{pt}], F).$$

Here $[M] \in H_n(M)$ denotes the fundamental class of M, and $[\mathrm{pt}] \in H_0(M)$ is the class of a point.

PROOF. We prove the claim for $\max F$, since the proof for $\min F$ is similar. Clearly,

$$c([M], F) \leq \max F.$$

Conversely, for any $\alpha < \max F$, the subset $\{F < \alpha\}$ is a noncompact n-manifold, and hence $H_n(\{F < \alpha\}) = 0$. In particular, $[M] \notin \mathrm{Image}(i_\alpha)$. Therefore $c([M], F) = \max F$. □

Exercise 10.1.3. Prove that $c(a, F) \leq c(a, G)$ for $F \leq G$, and that $c(a, F + \kappa) = c(a, F) + \kappa$ for $\kappa \in \mathbb{R}$. Deduce that

$$|c(a, F) - c(a, G)| \leq \max_M |F - G|.$$

In other words, $c(a, \cdot)$ is Lipschitz continuous with respect to the uniform norm on $C^\infty(M)$. Use this to extend the definition of $c(a, F)$ to continuous functions.

In the next proposition, we denote by $\cap$ the intersection product in homology. To recall its definition, see Section 10.4.

Proposition 10.1.4 (Subadditivity). *For any $F, G \in C^\infty(M)$ and any $a, b \in H_*(M) \setminus \{0\}$ we have*

$$c(a \cap b, F + G) \leq c(a, F) + c(b, G).$$

PROOF. Let $\varepsilon > 0$, and let α, β such that

$$c(a, F) < \alpha \leq c(a, F) + \varepsilon, \quad c(b, G) < \beta \leq c(b, G) + \varepsilon,$$

and

$$a \in \mathrm{Image}(i^F_\alpha), \quad b \in \mathrm{Image}(i^G_\beta),$$

where i^F_α and i^G_β denote the homomorphisms induced by the inclusions of sub-level sets of F and G, respectively. Let $A \subset \{F \leq \alpha\}$ and $B \subset \{G \leq \beta\}$ be piecewise smooth cycles such that

$$i^F_\alpha[A] = a, \quad i^G_\beta[B] = b.$$

Note that their intersection $A \cap B$ satisfies

$$A \cap B \subset \{F + G \leq \alpha + \beta\},$$

and by definition,

$$i_{\alpha+\beta}^{F+G}[A \cap B] = a \cap b.$$

Therefore, $c(a \cap b, F+G) \leq \alpha+\beta$. Since $\varepsilon > 0$ was arbitrary, the proposition follows. □

Proposition 10.1.5 (Poincaré duality). *For any $a \in H_*(M) \setminus \{0\}$,*

$$c(a, F) = - \inf_{b: b \circ a \neq 0} c(b, -F),$$

where $\circ$ stands for the intersection index modulo two.

Proof. Without loss of generality assume that F is a Morse function. First, suppose that $a \circ b \neq 0$. In particular, a and b are classes of complementary dimension, and so $a \cap b = (a \circ b)[\mathrm{pt}] \neq 0$. By the subadditivity,

$$0 = c(a \cap b, 0) = c\big(a \cap b, F + (-F)\big) \leq c(a, F) + c(b, -F).$$

This holds for all such b, and hence

$$c(a, F) \geq - \inf_{b: b \circ a \neq 0} c(b, -F).$$

Conversely, take $\varepsilon > 0$ such that the interval $\big(c(a, F) - \varepsilon, c(a, F)\big)$ contains no critical values of F. Take any α lying in this interval. Then $a \notin \mathrm{Image}(i_\alpha)$. Consider the long exact sequence of the pair $(M, \{F < \alpha\})$:

$$\cdots \to H_*(\{F < \alpha\}) \xrightarrow{i_\alpha} H_*(M) \xrightarrow{j} H_*(M, \{F < \alpha\}) \to \cdots$$

Then the assumption on α implies that $j(a) \neq 0 \in H_*(M, \{F < \alpha\})$. Now, Poincaré–Lefschetz duality gives an isomorphism

$$\mathrm{PD}\colon H_*(M, \{F < \alpha\}) \xrightarrow{\sim} H^{n-*}(\{F \geq \alpha\}).$$

Therefore, there is some

$$b_0 \in H_{n-*}(\{F \geq \alpha\}) = H_{n-*}(\{-F \leq -\alpha\})$$

such that $\langle \mathrm{PD}(a), b_0 \rangle \neq 0$. Let $b \in H_{n-*}(M)$ be the image of b_0 under

$$H_{n-*}(\{-F \leq -\alpha\}) \to H_{n-*}(M).$$

Then $c(b, -F) \leq -\alpha$ and, by construction, $a \circ b = \langle \mathrm{PD}(a), b_0 \rangle \neq 0$. Therefore,

$$\inf_{b: b \circ a \neq 0} c(b, -F) \leq -\alpha \leq -c(a, F) + \varepsilon.$$

Since ε was arbitrary, we get

$$c(a, F) \leq - \inf_{b: b \circ a \neq 0} c(b, -F).$$

This completes the proof. □

Exercise 10.1.6. Compute all the spectral numbers for the function $F(x, y) = \sin x \sin y$ on the torus $\mathbb{T}^2 = \mathbb{R}^2 / 2\pi\mathbb{Z}^2$.

Let us mention that spectral numbers are related to the following notion of algebraic flavor which, to the best of our knowledge, originated in dynamical systems. Let V be a finite-dimensional vector space over a field k. A function $\chi\colon V \to \mathbb{R} \cup \{-\infty\}$ is called a *characteristic exponent* if it satisfies the following axioms:

- $\chi(v) \in \mathbb{R}$ for all $v \neq 0$, and $\chi(0) = -\infty$.
- $\chi(\lambda v) = \chi(v)$ for all $\lambda \in k \setminus \{0\}$ and all $v \in V$.
- $\chi(v_1 + v_2) \leq \max\{\chi(v_1), \chi(v_2)\}$ for all $v_1, v_2 \in V$.

Exercise 10.1.7. A mainstream example (from the theory of dynamical systems) is the following: Let $A \in \mathrm{GL}(n, \mathbb{C})$. Prove that the function

$$\chi_A(v) = \lim_{n\to\infty} \frac{1}{n} \log|A^n v|$$

is a characteristic exponent on $\mathbb{C}^n$.

Now, let χ be a characteristic exponent on a vector space V.

Exercise 10.1.8. Prove that for every $\alpha \in \mathbb{R}$ the set $\{\chi < \alpha\}$ is a linear subspace of V.

Exercise 10.1.9. Prove that χ assumes at most n distinct real values, where $n = \dim_k V$.

Exercise 10.1.10. Prove that, for any $F \in C^\infty(M)$, the function

$$H_*(M) \to \mathbb{R} \cup \{-\infty\}, \quad a \mapsto c(a, F)$$

is a characteristic exponent (here we make the convention that $c(0, F) = -\infty$).

10.2. Morse homology

In this section we briefly sketch a Morse theoretical approach to the homology ring of a closed manifold M which serves as a prototype for Floer theory. Mathematics of the XXth century produced a number of remarkable applications of differential equations to topology of manifolds: Donaldson's invariants of four-manifolds based on instantons, Gromov's theory of pseudo-holomorphic curves in symplectic manifolds, and Seiberg–Witten theory, just to mention a few of them. Morse theory, which extracts basic topological information on manifolds from gradient ODEs, is perhaps the oldest, the simplest, and the most influential example of such an application[2]. We refer to the monographs [12, 14, 109, 143], as well as to the lecture notes [88] for a detailed exposition. To avoid technical issues of orientations, we will work with $\mathbb{Z}_2$-coefficients, although the construction below can be generalized to other fields.

[2]Unfortunately, this does not prevent certain mathematics departments from considering ODEs and PDEs as applied mathematics. Hopefully, a mentality shift is coming soon.

Recall that a critical point q of a smooth function $F\colon M \to \mathbb{R}$ is called *nondegenerate* if the Hessian $\partial^2 F/\partial x^2(q)$ is nondegenerate. In this case one can choose local coordinates $x_1, \ldots, x_n$ near q so that q corresponds to the origin and F is given by

$$F(x) = F(q) - x_1^2 - \cdots - x_k^2 + x_{k+1}^2 + \cdots + x_n^2. \tag{10.1}$$

The number k of negative indices is called the (Morse) index of the point q and is denoted by $\operatorname{ind} q$.

A function is called *Morse* if all of its critical points are nondegenerate. Morse functions are known to form an open dense set in the C^∞-topology on $C^\infty(M)$. We denote by $\operatorname{Crit}(F)$ the set of all critical points of a Morse function F, and by $\operatorname{Crit}_k(F)$ the subset of all critical points of index k.

Given a Morse function F and a Riemannian metric ρ on M, consider the negative gradient flow equation

$$\dot{u}(t) = -\nabla_\rho F\big(u(t)\big). \tag{10.2}$$

For critical points $x, y \in \operatorname{Crit}(F)$, consider the following *solution space*

$$\mathcal{M}(x,y) = \Big\{u\colon \mathbb{R} \to M : \dot{u} = -\nabla_\rho F(u),\ u(t) \xrightarrow[t\to-\infty]{} x,\ u(t) \xrightarrow[t\to\infty]{} y\Big\}.$$

In other words, $\mathcal{M}(x,y)$ is the set of all (parameterized) negative gradient trajectories connecting x to y.

Proposition 10.2.1. *For a "generic" pair* (F, ρ), $\mathcal{M}(x,y)$ *is a smooth manifold of dimension* $\operatorname{ind} x - \operatorname{ind} y$.

Let us comment on the genericity assumption. The *unstable* manifold of x, $W^{\mathrm{u}}(x)$, is the set of all points whose trajectories under the negative gradient flow converge to x in negative time. That is, if we denote by ϕ_t the flow of $-\nabla_\rho F$ (i.e., the solution of equation (10.2)),

$$W^{\mathrm{u}}(x) = \Big\{z \in M : \phi_t(z) \xrightarrow[t\to-\infty]{} x\Big\}.$$

Similarly, the *stable* manifold of y is defined by

$$W^{\mathrm{s}}(y) = \Big\{z \in M : \phi_t(z) \xrightarrow[t\to\infty]{} y\Big\}.$$

One can show that these are smooth manifolds of dimensions $\dim W^{\mathrm{u}}(x) = \operatorname{ind} x$ and $\dim W^{\mathrm{s}}(y) = n - \operatorname{ind} y$. The dimension count can be readily predicted with the help of the local normal form (10.1): Assume that the metric ρ is the Euclidean one. Then locally the negative gradient flow is given by

$$\begin{cases} \dot{u} = 2u, \\ \dot{v} = -2v, \end{cases}$$

where $u = (x_1, \ldots, x_k)$ and $v = (x_{k+1}, \ldots, x_n)$. It follows that $u(t) = \mathrm{e}^{2t}u(0)$ and $v(t) = \mathrm{e}^{-2t}v(0)$, so that $W^{\mathrm{u}}(0)$ is the k-dimensional subspace $\{v = 0\}$, while $W^{\mathrm{s}}(0)$ is the $(n-k)$-dimensional subspace $\{u = 0\}$. Note now that $\mathcal{M}(x,y)$ is precisely the intersection $W^{\mathrm{u}}(x) \cap W^{\mathrm{s}}(y)$, and for a generic pair

(F, ρ) this intersection is transversal for all $x, y \in \mathrm{Crit}(F)$, in which case Proposition 10.2.1 follows.

Since equation (10.2) is autonomous, the solution space $\mathcal{M}(x, y)$ (with $x \neq y$) admits a free $\mathbb{R}$-action by time translations:

$$s\colon u(t) \mapsto u(s+t).$$

In this context the group $\mathbb{R}$ is called *the gauge group*, and the quotient $\overline{\mathcal{M}}(x, y) := \mathcal{M}(x, y)/\mathbb{R}$ is called *the moduli space*. One can think of the moduli space as of the space of nonparametrized gradient trajectories connecting x and y. A basic dimension count shows that $\dim \overline{\mathcal{M}}(x, y) = \operatorname{ind} x - \operatorname{ind} y - 1$. Let us indicate two important particular cases:

- If $\operatorname{ind} x \leq \operatorname{ind} y$, then $\overline{\mathcal{M}}(x, y) = \varnothing$. That is, the index of critical points can only decrease along flow lines.
- If $\operatorname{ind} x - \operatorname{ind} y = 1$, then $\overline{\mathcal{M}}(x, y)$ is a zero-dimensional manifold.

Example 10.2.2. Consider $S^1 = \{(u, v) \in \mathbb{R}^2 : u^2 + v^2 = 1\}$, with the height function $F(u, v) = v$. The function F has two critical points, a maximum at $x = (0, 1)$, and a minimum at $y = (0, -1)$. Then $\operatorname{ind} x = 1$ and $\operatorname{ind} y = 0$, and hence $\dim \overline{\mathcal{M}}(x, y) = 0$. As can be seen in Figure 10.1, it consists of two points.

In general, the manifold $\overline{\mathcal{M}}(x, y)$ is not compact. Let us illustrate the way in which compactness fails in the following example.

Example 10.2.3. Consider $\mathbb{T}^2 = S^1 \times S^1$, and let F be the sum of the height functions on each factor. It has unique global maximum and minimum and two saddle points. Denote these x, y, z_1 and z_2, respectively. Then

$$\operatorname{ind} x = 2, \quad \operatorname{ind} y = 0, \quad \operatorname{ind} z_1 = \operatorname{ind} z_2 = 1.$$

Since $\operatorname{ind} x - \operatorname{ind} y = 2$, we have $\dim \overline{\mathcal{M}}(x, y) = 1$. Figure 10.2 shows the one-parameter family of flow lines from x to y. Observe that this family is not compact—intuitively, one can see that going to the left, say, the flow lines converge to a 'broken trajectory,' i.e., the union of flow lines from x to z_1 and from z_1 to y.

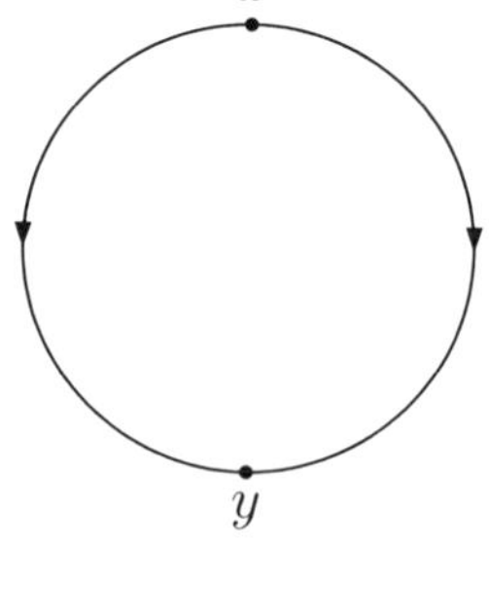

FIGURE 10.1. Negative gradient trajectories for the height function

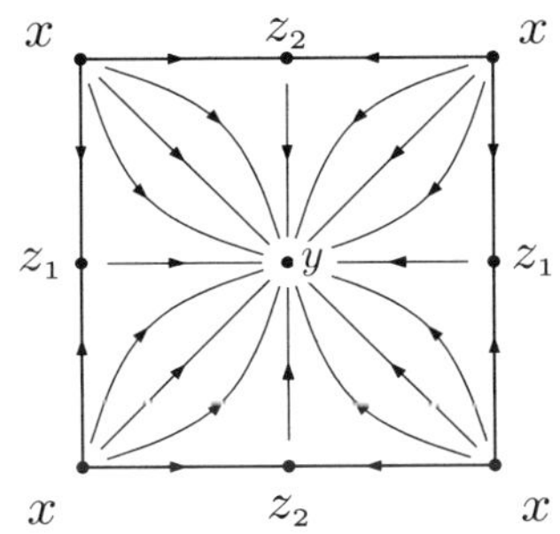

FIGURE 10.2. Compactness breaking in Morse homology

This is a general phenomenon. In general, $\overline{\mathcal{M}}(x, y)$ can be compactified by adding broken trajectories. More precisely, it admits a natural compactification $\widetilde{\mathcal{M}}(x, y)$, which is a stratified manifold, whose stratum of codimension k consists of k-times broken trajectories, i.e., flow lines connecting k intermediate critical points. Two important cases are:

(1) If $\operatorname{ind} x - \operatorname{ind} y = 1$, then there are no critical points z with $\operatorname{ind} x < \operatorname{ind} z < \operatorname{ind} y$. Thus, the manifold $\overline{\mathcal{M}}(x, y)$ is already a compact zero-dimensional manifold, that is, a finite collection of points. In this case we write

$$n(x, y) = \#_2 \overline{\mathcal{M}}(x, y) \tag{10.3}$$

for the number of these points taken modulo 2.

(2) If $\operatorname{ind} x - \operatorname{ind} y = 2$, $\widetilde{\mathcal{M}}(x, y)$ is a compact one-dimensional manifold whose boundary is

$$\partial \widetilde{\mathcal{M}}(x, y) = \bigcup_{\substack{z \in \operatorname{Crit}(F) \\ \operatorname{ind} z = \operatorname{ind} y + 1}} \overline{\mathcal{M}}(x, z) \times \overline{\mathcal{M}}(z, y).$$

Since the number of boundary components of a compact one-dimensional manifold is even, we have

$$\begin{aligned} 0 = \# \, \partial \widetilde{\mathcal{M}}(x, y) &= \sum_{\substack{z \in \operatorname{Crit}(F) \\ \operatorname{ind} z = \operatorname{ind} y + 1}} \# \, \overline{\mathcal{M}}(x, z) \cdot \# \, \overline{\mathcal{M}}(z, y) \\ &= \sum_{\substack{z \in \operatorname{Crit}(F) \\ \operatorname{ind} z = \operatorname{ind} y + 1}} n(x, z) \cdot n(z, y). \end{aligned}$$

Now we are ready to define Morse homology. Introduce the *Morse chain complex*

$$\mathrm{CM}_k(F) = \operatorname{span}_{\mathbb{Z}_2} \operatorname{Crit}_k(F),$$

equipped with the differential

$$\mathrm{d} = \mathrm{d}_\rho \colon \mathrm{CM}_k(F) \to \mathrm{CM}_{k-1}(F), \quad \mathrm{d}x = \sum_{z \in \operatorname{Crit}_{k-1}(F)} n(x, z) \cdot z.$$

Note that the definition of d depends on the metric ρ. Let us check that this is indeed a complex, that is $\mathrm{d}^2 = 0$. Take any $x \in \mathrm{Crit}_k(F)$. Then

$$\begin{aligned}\mathrm{d}^2 x &= \sum_{z \in \mathrm{Crit}_{k-1}(F)} n(x,z)\,\mathrm{d}z = \sum_{z \in \mathrm{Crit}_{k-1}(F)} n(x,z) \sum_{y \in \mathrm{Crit}_{k-2}(F)} n(z,y) \cdot y \\ &= \sum_{y \in \mathrm{Crit}_{k-2}(F)} y \cdot \sum_{z \in \mathrm{Crit}_{k-1}(F)} n(x,z) \cdot n(z,y) = 0,\end{aligned}$$

where the last equality follows from (10.4).

The homology of the complex $(\mathrm{CM}(F), \mathrm{d})$ is called the *Morse homology* of the pair (F, ρ), and we denote it by $\mathrm{HM}_*(F, \rho)$.

Example 10.2.4. Consider again the circle S^1 with a height function F, as in Example 10.2.2. Then the Morse complex consists of $\mathrm{CM}_1(F) = \mathbb{Z}_2 \cdot x$ and $\mathrm{CM}_0(F) = \mathbb{Z}_2 \cdot y$. We compute:

$$\mathrm{d}x = 2y = 0 \pmod 2, \quad \mathrm{d}y = 0,$$

and so

$$\mathrm{HM}_k(F) = \begin{cases} \mathbb{Z}_2, & k = 0, 1, \\ 0, & k \geq 2. \end{cases}$$

10.3. Canonical morphisms

The Morse complexes we have defined clearly depend on the pair (F, ρ). Their homologies, however, are canonically isomorphic. Let $(F^{(1)}, \rho^{(1)})$ and $(F^{(2)}, \rho^{(2)})$ be two generic pairs. Choose a generic pair of families $\{(F_t, \rho_t)\}_{t \in \mathbb{R}}$ such that

$$(F_t, \rho_t) = \begin{cases} (F^{(1)}, \rho^{(1)}), & t \leq -1, \\ (F^{(2)}, \rho^{(2)}), & t \geq 1. \end{cases}$$

Consider the ODE

$$\dot{u}(t) = -\nabla_{\rho_t} F_t\big(u(t)\big). \tag{10.4}$$

Note that this equation is nonautonomous, for the vector field $\nabla_{\rho_t} F_t$ depends on the time parameter. Therefore, the moduli space of solutions does not admit an $\mathbb{R}$-action in this case. It turns out that for $x \in \mathrm{Crit}(F^{(1)})$ and $y \in \mathrm{Crit}(F^{(2)})$ of equal indices, the space of solutions of (10.4) connecting x and y is a compact zero-dimensional manifold, and we denote by $\nu(x, y)$ the number (mod 2) of such trajectories. Define the *continuation map*:

$$\Phi \colon \mathrm{CM}(F^{(1)}) \to \mathrm{CM}(F^{(2)}), \quad x \mapsto \sum_y \nu(x, y) y.$$

Proposition 10.3.1. *The continuation maps enjoy the following properties*:

(1) *Φ is a chain map, and hence induces a map*

$$\Phi_* \colon \mathrm{HM}_*(F^{(1)}, \rho^{(1)}) \to \mathrm{HM}_*(F^{(2)}, \rho^{(2)}).$$

(2) *The induced map on homology,* Φ_**, does not depend on the choice of families* (F_t, ρ_t).

(3) *Given three pairs* $(F^{(i)}, \rho^{(i)})$, $i = 1, 2, 3$ *and continuation maps*

$$\Phi^{(ij)}\colon \mathrm{CM}(F^{(i)}) \to \mathrm{CM}(F^{(j)}) \quad \text{for } i < j,$$

the following diagram commutes:

$$\begin{array}{ccc} \mathrm{HM}_*(F^{(1)}) & \xrightarrow{\Phi_*^{(12)}} & \mathrm{HM}_*(F^{(2)}) \\ & \Phi_*^{(13)} \searrow \quad \swarrow \Phi_*^{(23)} & \\ & \mathrm{HM}_*(F^{(3)}) & \end{array}$$

Now we can easily deduce the invariance of Morse homology. Note that, by taking the constant families $(F_t, \rho_t) \equiv (F, \rho)$, we get a continuation map $\Phi\colon \mathrm{CM}(F^{(1)}) \to \mathrm{CM}(F^{(1)})$ such that $\Phi_* = \mathbb{1}$. Now, given two pairs $(F^{(i)}, \rho^{(i)})$, $i = 1, 2$, we get, using this fact and Proposition 10.3.1 that $\Phi_*^{(12)} \circ \Phi_*^{(21)} = \mathbb{1}$ and $\Phi_*^{(21)} \circ \Phi_*^{(12)} = \mathbb{1}$. Therefore,

$$\mathrm{HM}_*(F^{(1)}, \rho^{(1)}) \simeq \mathrm{HM}_*(F^{(2)}, \rho^{(2)})$$

canonically. Identifying via the canonical morphisms the vector spaces $\mathrm{HM}_*(F, \rho)$, where (F, ρ) runs over all generic pairs, we get a vector space $\mathrm{HM}_*(M)$ which is called *the Morse homology* of M.

It turns out that the Morse homology is canonically isomorphic to the singular homology with $\mathbb{Z}_2$-coefficients: $\mathrm{HM}_*(M) \simeq H_*(M; \mathbb{Z}_2)$. Let us describe this isomorphism (we follow here the presentation in [88]). Denote by $C_*^M = \mathrm{CM}_*(F)$ the Morse chain complex associated to a generic pair consisting of a Morse function F and a Riemannian metric ρ, and by C_*^S the singular chain complex. Define chain maps

$$D\colon C_*^M \to C_*^S, \quad A\colon C_*^S \to C_*^M$$

as follows. For $x \in \mathrm{Crit}(F)$ set

$$D(x) = \mathrm{Closure}\big(W^{\mathrm{u}}(x)\big). \tag{10.5}$$

Next, for a piece-wise smooth simplex Δ and $z \in \mathrm{Crit}(F)$ denote by $\mathcal{M}(\Delta, z)$ the space of all points lying on negative gradient trajectories from Δ to z. In other words, $\mathcal{M}(\Delta, z) = \Delta \cap W^{\mathrm{s}}(z)$. Note that $\mathcal{M}(\Delta, z)$ is zero-dimensional exactly when $\dim \Delta = \mathrm{ind}\, z$. Define

$$A(\Delta) = \sum_z \#_2 \mathcal{M}(\Delta, z) \cdot z \tag{10.6}$$

where, as before, $\#_2 \mathcal{M}(\Delta, z) = 0$ unless $\dim \mathcal{M}(\Delta, z) = 0$.

Write ϕ_s for the negative gradient flow.

Exercise 10.3.2. (i) Show that $AD = \mathbb{1}$.

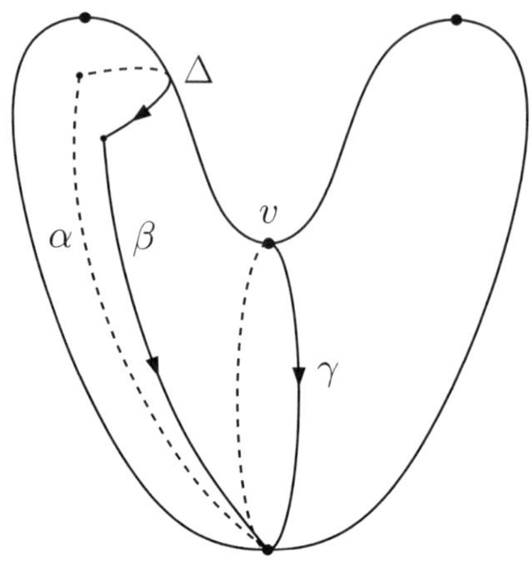

FIGURE 10.3. DA is chain homotopic to the identity

(ii) Define a map

$$\Phi\colon C_i^S \to C_{i+1}^S, \quad \Delta \mapsto \mathrm{Closure}\Big(\bigcup_{s\geq 0} \phi^s \Delta\Big).$$

Show that Φ is a chain homotopy between DA and $\mathbb{1}$:

$$DA - \mathbb{1} = \partial\Phi + \Phi\partial.$$

This is illustrated in Figure 10.3, where Δ is an interval on the heart-shaped sphere. In this case $A(\Delta) = v$ and hence $DA(\Delta) = D(v) = \gamma$. On the other hand, $\partial\Phi(\Delta) = \Delta + \alpha + \beta + \gamma$, while $\Phi\partial(\Delta) = \alpha + \beta$. As we are working modulo 2, we get

$$(\partial\Phi + \Phi\partial)(\Delta) = \Delta + \gamma = (DA - \mathbb{1})(\Delta).$$

It follows from the exercise that D and A induce mutually inverse isomorphisms between the Morse homology and singular homology.

10.4. Ring structure

Singular homology $H_*(M;\mathbb{Z}_2)$ carries a natural ring structure with respect to the intersection product. Recall that the latter is defined as follows. Let $a, b \in H_*(M;\mathbb{Z}_2)$, and assume without loss of generality that they are represented by piecewise smooth cycles $A, B \subset M$, respectively, intersecting transversally. Then their intersection $A \cap B$ is again a piecewise smooth cycle, and one defines $a \cap b = [A \cap B]$. This defines

$$\cap : H_{n-k}(M;\mathbb{Z}_2) \otimes H_{n-l}(M;\mathbb{Z}_2) \to H_{n-k-l}(M;\mathbb{Z}_2).$$

Let us describe the counterpart of this product in Morse homology. Take any generic triple (F_i, ρ_i), $i = 1, 2, 3$. Pick critical points $x_i \in \mathrm{Crit}(F_i)$. Denote $\nabla_i := \nabla_{\rho_i}$, and write ϕ_t^i for the flow of $-\nabla_i F_i$. Consider the space of solutions $\mathcal{M}(x_1, x_2, x_3)$ of the following problem:

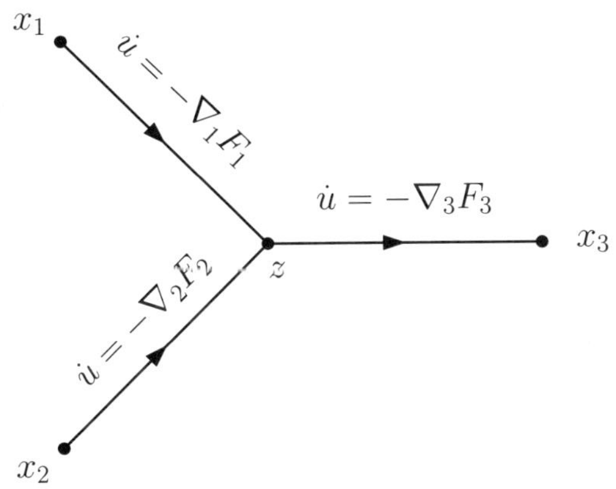

FIGURE 10.4. Product in Morse homology

More precisely,

$$\begin{aligned}\mathcal{M}(x_1, x_2, x_3) &= \Big\{ z \in M : \phi_t^1(z) \xrightarrow[t\to-\infty]{} x_1, \phi_t^2(z) \xrightarrow[t\to-\infty]{} x_2, \phi_t^3(z) \xrightarrow[t\to\infty]{} x_3 \Big\} \\ &= W^{\mathrm{u}}_{\rho_1}(x_1) \cap W^{\mathrm{u}}_{\rho_2}(x_2) \cap W^{\mathrm{s}}_{\rho_3}(x_3).\end{aligned}$$

From the last description it is clear that, when the intersection is transversal, the manifold $\mathcal{M}(x_1, x_2, x_3)$ is zero-dimensional provided

$$\operatorname{ind} x_1 + \operatorname{ind} x_2 - n = \operatorname{ind} x_3.$$

We define a product in Morse homology by

$$x_1 \otimes x_2 \mapsto x_1 \cap_M x_2 = \sum_{x_3} \#_2\, \mathcal{M}(x_1, x_2, x_3) \cdot x_3,$$

where by convention, $\#_2\mathcal{M}(x_1, x_2, x_3) = 0$ unless $\dim \mathcal{M}(x_1, x_2, x_3) = 0$. In order to see that this product corresponds to the intersection product in singular homology, let us apply the morphisms D and A defined by (10.5) and (10.6): Let (F_i, ρ_i), $i = 1, 2, 3$ be Morse pairs, and denote by A_i, D_i the corresponding chain maps. Let $\alpha_i = D_i(x_i)$ for $x_i \in \operatorname{Crit}(F_i)$. Then

$$\alpha_1 \cap \alpha_2 = \big[\mathrm{Closure}\big(W^{\mathrm{u}}(x_1)\big) \cap \mathrm{Closure}\big(W_2^{\mathrm{u}}(x_2)\big)\big],$$

and for any $x_3 \in \operatorname{Crit}(F_3)$

$$\mathcal{M}(\alpha_1 \cap \alpha_2, x_3) = \alpha_1 \cap \alpha_2 \cap W_3^{\mathrm{s}}(x_3).$$

In generic situations triple intersection points lie away from the boundaries of α_1 and α_2, and therefore

$$\mathcal{M}(\alpha_1 \cap \alpha_2, x_3) = W_1^{\mathrm{u}}(x_1) \cap W_2^{\mathrm{u}}(x_2) \cap W_3^{\mathrm{s}}(x_3) = \mathcal{M}(x_1, x_2, x_3).$$

Therefore,

$$A_3\big(D_1(x_1) \cap D_2(x_2)\big) = \sum_{x_3} \#_2\, \mathcal{M}(x_1, x_2, x_3) \cdot x_3 = x_1 \cap_M x_2.$$

We conclude that A and D induce isomorphisms *of rings.*

10.5. Morse–Novikov homology

As noted above, our description of Morse homology is meant to serve as an introduction to Floer homology. However, in Floer homology we encounter the problem that the functional we wish to study is not necessarily single-valued, and so we need to pass to a suitable covering space. On finite-dimensional manifolds this problem is settled in the framework of *Morse–Novikov* homology of closed forms (see, e.g., [140])

Let M be a closed manifold, and α a closed 1-form on M. The form α determines a homomorphism $I_\alpha\colon \pi_1(M) \to \mathbb{R}$ by $[\gamma] \mapsto \int_\gamma \alpha$. Assume that its image is a discrete subgroup of $\mathbb{R}$, i.e.,

$$I_\alpha\big(\pi_1(M)\big) = B \cdot \mathbb{Z}, \quad B > 0.$$

Denote $\overline{\pi}_1(M) = \pi_1(M)/\ker I_\alpha \simeq \mathbb{Z}$. Let $S \in \overline{\pi}_1(M)$ be a generator with $I_\alpha(S) = B$. Define a cover $\widetilde{M} \to M$ as follows. Fix a base point $z_* \in M$ and let $\widetilde{M}$ be the set of equivalence classes of pairs (x, γ), where $x \in M$, γ is a path from z_* to x, and

$$(x, \gamma) \sim (x', \gamma') \iff x = x',\ \gamma \# (-\gamma') \in \ker I_\alpha,$$

where $\#$ denotes concatenation of paths. For brevity, we will denote the equivalence class $[(x, \gamma)]$ by (x, γ). Note that the group of deck transformations of $\widetilde{M} \to M$ is naturally isomorphic to $\overline{\pi}_1(M)$, and denote this action, which is illustrated in Figure 10.5, by

$$S \cdot (x, \gamma) = (x, S\#\gamma). \tag{10.7}$$

On $\widetilde{M}$ we can define a smooth function F by

$$F(x, \gamma) = \int_\gamma \alpha.$$

Note that $\mathrm{d}F = \alpha$. Thus we can define the index of a critical point x of α as the index of any lift $\tilde{x}$ of x, as a critical point of F. Now, choose a metric ρ on M and lift it to $\tilde{\rho}$ on $\widetilde{M}$. Using the metric $\tilde{\rho}$ we define the negative gradient flow ϕ_t of F on $\widetilde{M}$. Denote by $n(\tilde{x}, \tilde{y})$ the number (modulo 2) of trajectories connecting $\tilde{x}$ and $\tilde{y}$, for $\tilde{x}$ and $\tilde{y}$ such that $\operatorname{ind}\tilde{x} - \operatorname{ind}\tilde{y} = 1$ (see formula (10.3) above).

Since F has infinitely many critical points, we need a way to keep track of different lifts of a critical point $x \in M$. To this end we introduce the field

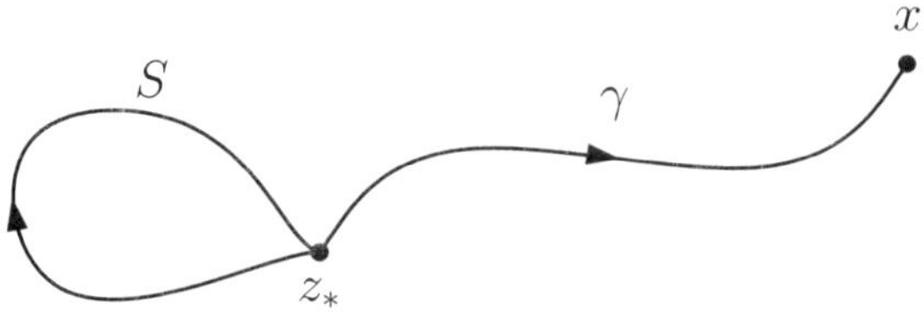

FIGURE 10.5. Action of $\overline{\pi}_1(M)$ on $\widetilde{M}$

of Laurent series:

$$\mathcal{K} = \mathbb{Z}_2[[s]] = \Big\{ \sum_{i \in \mathbb{Z}} a_i s^i : a_i \in \mathbb{Z}_2, a_i = 0 \text{ for all } i > N, \text{ for some } N \in \mathbb{Z} \Big\}.$$

Exercise 10.5.1. Prove that $\mathcal{K}$ is a field.

Denote by $\mathrm{Crit}_i(F)$ the set of index i critical points of F on $\widetilde{M}$. Define $\mathrm{CN}_i(\alpha)$ to be the space of all finite linear combinations of elements of $\mathrm{Crit}_i(F)$ with coefficients in $\mathcal{K}$, where we identify $s \cdot (x, \gamma)$ with $S \cdot (x, \gamma)$ (defined by (10.7)), and consider it as a vector space over $\mathcal{K}$. Define the differential $\mathrm{d}\colon \mathrm{CN}_i(\alpha) \to \mathrm{CN}_{i-1}(\alpha)$ by

$$\mathrm{d}\tilde{x} = \sum_{\tilde{y} \in \mathrm{Crit}_{i-1}(F)} n(\tilde{x}, \tilde{y}) \tilde{y}. \tag{10.8}$$

Let us check that d is well defined, that is, $\mathrm{d}\tilde{x}$ indeed lies in $\mathrm{CN}_{i-1}(\alpha)$. First we choose a basis in $\mathrm{CN}_i(\alpha)$ and $\mathrm{CN}_{i-1}(\alpha)$ as follows: Let $x_1, \ldots, x_N$ and $y_1, \ldots, y_L$ be all critical points of α of indices i and $i-1$, respectively. Choose their lifts $\tilde{x}_1, \ldots, \tilde{x}_N$ and $\tilde{y}_1, \ldots, \tilde{y}_L$ to $\widetilde{M}$, a unique lift for each critical point. With this notation

$$\mathrm{d}\tilde{x}_i = \sum_{j=1}^{L} \Big(\sum_{k \in \mathbb{Z}} n(\tilde{x}_i, S^k \tilde{y}_j) s^k \Big) \cdot \tilde{y}_j. \tag{10.9}$$

We have to check that the coefficient at $\tilde{y}_j$ lies in $\mathcal{K}$, that is, it satisfies the finiteness condition. Let us make two observations:

(1) $F(S^k \tilde{x}) = F(\tilde{x}) + kB$.
(2) F decreases along the trajectories of the negative gradient flow: $F\big(\phi_t(\tilde{x})\big) < F(\tilde{x})$ for $t > 0$. In particular, if there are trajectories connecting $\tilde{x}$ to $\tilde{y}$, then $F(\tilde{y}) < F(\tilde{x})$.

Now, assume that $n(\tilde{x}_i, S^k \tilde{y}_j) \neq 0$. Then by 1 and 2,

$$F(\tilde{x}_i) > F(S^k \tilde{y}_j) = F(\tilde{y}_j) + kB \implies k < \frac{F(\tilde{x}_i) - F(\tilde{y}_j)}{B}.$$

In particular, $n(\tilde{x}_i, S^k \tilde{y}_j) = 0$ for large enough k, and thus the right-hand side of (10.9) is a well defined element in $\mathrm{CN}_{i-1}(\alpha)$.

Similarly to the case of Morse homology, we find that $\mathrm{d}^2 = 0$. The homology of the complex $(\mathrm{CN}(\alpha), \mathrm{d})$ is called the Morse–Novikov homology of α.

Theorem 10.5.2. *The Morse–Novikov homology of α depends only on the cohomology class $[\alpha] \in H^1(M; \mathbb{R})$.*

Example 10.5.3. Consider the case $M = S^1$, with coordinate t. Let $H\colon S^1 \to \mathbb{R}$ be the height function. Let $\alpha = \mathrm{d}H + \varepsilon\, \mathrm{d}t$ for some sufficiently small $\epsilon > 0$. Then $I_\alpha\big(\pi_1(S^1)\big) = \varepsilon \cdot \mathbb{Z}$, and $\widetilde{M}$ is the covering of S^1 with deck group $\mathbb{Z}$, i.e. the universal covering $\mathbb{R} \to S^1$. The Novikov complex of α is a 'small perturbation' of that of H. So, denote by x and y the

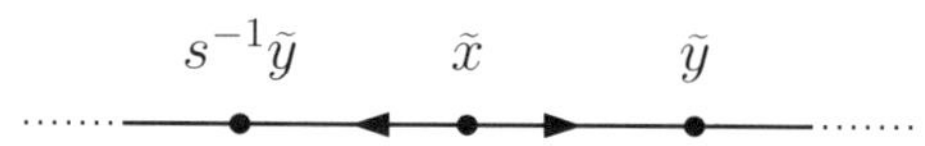

FIGURE 10.6. The negative gradient flow on $\widetilde{M} = \mathbb{R}$

unique maximum and minimum, respectively, of H on S^1, and choose lifts $\tilde{x}$, $\tilde{y}$ as described in Figure 10.6. Then $\mathrm{d}\tilde{y} = 0$ for index reasons, and $\mathrm{d}\tilde{x} = \tilde{y} + s^{-1}\tilde{y} = (1+s^{-1})\tilde{y}$ (recall that we are working modulo 2). Therefore, $\tilde{y} = \mathrm{d}(1+s^{-1})^{-1}\tilde{x}$ and hence the Morse–Novikov homology vanishes. This also follows from Theorem 10.5.2, since $[\alpha] = \varepsilon[\mathrm{d}t]$, and $\mathrm{d}t$ has no critical points.

CHAPTER 11

An Overview of Floer Theory

In this chapter we present an brief survey of Floer theory, an infinite-dimensional analogue of Morse–Novikov theory for the classical action functional associated to a 1-periodic Hamiltonian on a symplectic manifold. Here we outline the construction of Floer homology for *spherically monotone* symplectic manifolds. Floer theory, in this case, is as rich as in the general one, but the technicalities are significantly less involved. We refer the reader to the books [12, 108] for a detailed exposition.

11.1. Spherically monotone symplectic manifolds

Let (M, ω) be a closed symplectic manifold. Let

$$[\omega]\colon \pi_2(M) \to \mathbb{R}, \quad A \mapsto ([\omega], A) := \int_A \omega,$$

be the *symplectic area functional*, and let c_1 be the first Chern class of M (recall from Section 1.6 that this is the first Chern class of the tangent bundle TM, when the latter is equipped with some compatible almost complex structure) considered as a functional on $\pi_2(A)$.

Definition 11.1.1. A symplectic manifold (M, ω) is called (*spherically*) *monotone* if $\pi_2(M) \neq 0$, $[\omega] \neq 0$ and

$$[\omega] = \kappa c_1$$

for some $\kappa > 0$.

Examples of spherically monotone manifolds include, for instance, complex projective spaces and certain products thereof, as follows from the following exercise:

Exercise 11.1.2. Consider the complex projective space $\mathbb{C}P^n$ with the Fubini–Studi form ω_{FS}, defined in Section 1.7 above. Take the complex line

$$L = \{[z_0 : z_1 : 0 : \ldots : 0]\} \subset \mathbb{C}P^n.$$

Prove that

$$\big(c_1(\mathbb{C}P^n), [L]\big) = n + 1.$$

Deduce that $(\mathbb{C}P^n, \omega_{FS})$ is monotone, and more generally that $(\mathbb{C}P^n \times \mathbb{C}P^m, \alpha\omega_{FS} \oplus \beta\omega_{FS})$ is monotone if and only if $\alpha/\beta = (n + 1)/(m + 1)$. *Hint*: denote by H the tautological line bundle over $\mathbb{C}P^n$. Prove that the direct sum of the tangent bundle of $\mathbb{C}P^n$ with the trivial complex line bundle is isomorphic to the $(n + 1)$-fold direct sum $H^* \oplus \cdots \oplus H^*$.

For the remainder of this chapter, we assume that (M, ω) is a closed monotone symplectic manifold.

11.2. The least action principle

Let us recall some preliminaries on the symplectic action functional, which was introduced in Section 4.2. We repeat the constructions of Section 4.2 using slightly different notations. Denote $\overline{\pi}_2(M) = \pi_2(M)/\ker([\omega])$.[1] Since (M, ω) is monotone, this group necessarily has the form $\Omega \cdot \mathbb{Z}$ for some $\Omega > 0$. Clearly, $[\omega]$ descends to $\overline{\pi}_2(M)$.

Let ΛM be the space of all contractible loops $z\colon S^1 \to M$. Consider the following cover of ΛM:

$$\widetilde{\Lambda M} = \{(z, u) : z \in \Lambda, u\colon D^2 \to M, u|_{\partial D^2} = z\}/\sim$$

where

$$(z, u) \sim (z', u') \iff z = z',\ u \# (-u') = 0 \in \overline{\pi}_2(M).$$

Here # stands for gluing the discs along the boundary. For the sake of brevity, we shall often write (z, u) for the equivalence class $[(z, u)]$.

Let S be a generator of $\overline{\pi}_2(M)$ with $([\omega], S) = \Omega$, and denote $(c_1, S) =: N$. The number N is called the *minimal Chern number* of M. Note that $\Omega = \kappa N$, where κ is the constant in Definition 11.1.1 (so, in particular, $N > 0$).

The group of deck transformations of $\widetilde{\Lambda M} \to \Lambda$ is naturally isomorphic to $\overline{\pi}_2(M)$, and we denote the corresponding action by

$$S \cdot (z, u) = \big(z, u \# (-S)\big). \tag{11.1}$$

The action is illustrated in Figure 11.1.

Let $F(x, t)$ be a normalized Hamiltonian function, which we assume to be 1-periodic in time: $F(x, t+1) = F(x, t)$. Let f_t, $t \in [0, 1]$ be the Hamiltonian flow generated by F. Define the *action functional* $\mathcal{A}_F\colon \widetilde{\Lambda M} \to \mathbb{R}$ by

$$\mathcal{A}_F(z, u) = \int_0^1 F(z(t), t)\, \mathrm{d}t - \int_{D^2} u^*\omega.$$

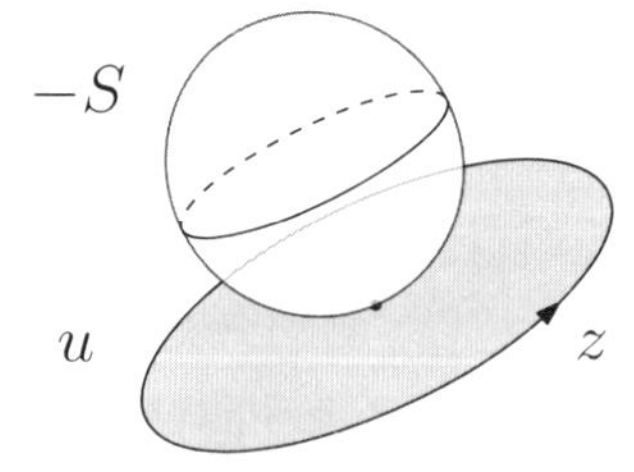

FIGURE 11.1. Action of $\overline{\pi}_2(M)$ on $\widetilde{\Lambda M}$

[1] In the notations of Section 4.2, $\overline{\pi}_2(M)$ is isomorphic to Γ_ω.

Exercise 11.2.1. Deduce from (11.1) that the action of S increases the value of $\mathcal{A}_F$ (cf. Section 10.5):

$$\mathcal{A}_F\big(S\cdot(z,u)\big) = \mathcal{A}_F(z,u) + \Omega.$$

The tangent space $T_{\tilde{z}}\widetilde{\Lambda M}$ at a point $\tilde{z} = (z,u)$ can be naturally identified with the space of tangent vector fields $\xi(t) \in T_{z(t)M}$ to M along the loop $z(t)$. Let us calculate the differential $\mathrm{d}\mathcal{A}_F(\xi)$ at a point $\tilde{z}$:

Exercise 11.2.2. Prove that

$$\mathrm{d}\mathcal{A}_F(\xi) = \int_0^1 \Big(\mathrm{d}F_t(\xi) - i_\xi\omega\big(\dot{z}(t)\big)\Big)\,\mathrm{d}t. \tag{11.2}$$

Recalling that the Hamiltonian vector field $\operatorname{sgrad} F_t$ is defined by $i_{\operatorname{sgrad} F_t}\omega = -\,\mathrm{d}F_t$ we rewrite this as

$$\mathrm{d}\mathcal{A}_F(\xi) = \int_0^1 \omega(\xi, -\dot{z}(t) + \operatorname{sgrad} F_t)\,\mathrm{d}t. \tag{11.3}$$

As an immediate consequence we get the following classical result.

Proposition 11.2.3 (The least action principle). *An element* $(z,u) \in \widetilde{\Lambda M}$ *is a critical point of* $\mathcal{A}_F$ *if and only if* z *is a contractible* 1*-periodic orbit of the Hamiltonian flow of* F.

Indeed, by (11.2), $\mathrm{d}_{(z,u)}\mathcal{A}_F(\xi) = 0$ for all ξ if and only if $\dot{z} = \operatorname{sgrad} F_t$.

11.3. The Floer equation

Our next goal is to calculate the gradient of $\mathcal{A}_F$ with respect to a special class of Riemannian metrics on $\widetilde{\Lambda M}$.

Choose a loop $J(t)$ of compatible almost complex structures and define a Riemannian metric ρ_J on $\widetilde{\Lambda M}$ as follows: Given $\xi(t), \eta(t) \in T_{\tilde{z}}\widetilde{\Lambda M}$, set

$$\langle \xi(t), \eta(t)\rangle_{\rho_J} = \int_0^1 \omega\big(\xi(t), J(t)\eta(t)\big)\,\mathrm{d}t.$$

By the definition of the gradient at a point $\tilde{z} = (z,u) \in \widetilde{\Lambda M}$,

$$\mathrm{d}\mathcal{A}_F(\xi) = \int_0^1 \omega(\xi(t), J(t)\nabla_{\rho_J}\mathcal{A}_F)\,\mathrm{d}t$$

for every vector field ξ along z. Comparing this with formula (11.3) above we get that

$$J(t)\nabla_{\rho_J}\mathcal{A}_F = -\dot{z}(t) + \operatorname{sgrad} F_t.$$

Multiplying by $J(t)$ and applying Exercise 1.6.1 we conclude that

$$\nabla_{\rho_J}\mathcal{A}_F(z,u) = J(t)\dot{z}(t) + \nabla_{J(t)}F_t\big(z(t)\big), \tag{11.4}$$

where $\nabla_{J(t)}$ denotes the gradient with respect to the Riemannian metric induced by $J(t)$ on M. The negative gradient flow of $\mathcal{A}_F$ is given by the equation

$$u\colon \mathbb{R} \to \widetilde{\Lambda M}, \quad \frac{\partial u}{\partial \tau} = -\nabla_{\rho_J}\mathcal{A}_F(u).$$

In view of (11.4) it can be interpreted as a partial differential equation

$$\partial_\tau u + J(t)\partial_t u + \nabla_{J(t)} F(u,t) = 0 \tag{11.5}$$

on M, for $u\colon \mathbb{R}(\tau) \times S^1(t) \to M$. We refer to (11.5) as *Floer's equation.* Observe that when $J(t)$ is independent of t, the first two terms can be interpreted as a nonlinear Cauchy-Riemann operator, and the last term as a nonhomogeneous correction.

Remark 11.3.1. In contrast to the finite-dimensional situation, the negative gradient equation of the action functional *does not* define a flow on the loop space. To see this, consider the simplest local model when $M = \mathbb{C}$, $J(t)$ is independent of t and coincides with the standard complex structure $\xi \mapsto \mathrm{i}\xi$, and the nonhomogeneous term vanishes: $F = 0$. In this situation we can ignore spanning discs and deal with smooth loops $z\colon S^1 \to \mathbb{C}$. The flow in question is determined by the Cauchy problem

$$\partial_\tau u = -\mathrm{i}\partial_t u, \quad u\colon \mathbb{R}(\tau) \times S^1(t) \to \mathbb{C}, \quad u(0,t) = z(t). \tag{11.6}$$

The solution $u(\tau,t)$ is interpreted as the evolution of the initial condition $z \in \Lambda\mathbb{C}$ for time τ. Let us expand $u(\tau,t)$ in a Fourier series with respect to t:

$$u(\tau,t) = \sum_{k\in\mathbb{Z}} a_k(\tau)\mathrm{e}^{2\pi \mathrm{i} kt}. \tag{11.7}$$

Assume that the Fourier coefficients of the initial loop $u(0,t) = z(t)$ satisfy

$$b_1 \mathrm{e}^{-|k|\epsilon} \le |a_k(0)| \le b_2 \mathrm{e}^{-|k|\epsilon}$$

for some $b_2 > b_1 > 0$ and $\epsilon > 0$. The upper bound on the Fourier coefficients guarantees that the initial condition $z(t)$ is C^∞-smooth. Then (11.6) yields $\dot a_k = 2\pi k a_k$, and hence $a_k(\tau) = \mathrm{e}^{2\pi k\tau} a_k(0)$. Therefore

$$|a_k(\tau)| \ge b_1 \exp(-|k|\epsilon + k\tau),$$

and thus for $\tau > \epsilon$ the Fourier coefficients in (11.7) go to infinity as $k \to +\infty$. In particular, the evolution of z is not defined at the time moment $\tau = \epsilon$. Since ϵ can be chosen arbitrarily close to zero, we conclude that equation (11.6) does not define an evolution on the loop space.

A crucial finding of Floer was that even though the negative gradient flow of the action functional does not give rise to a dynamics on the loop space, the gradient trajectories connecting pairs of critical points are well defined. These trajectories are the solutions of the *boundary value problem* which we are just going to present. Denote by $\mathcal{P}(F)$ the set of all contractible 1-periodic Hamiltonian orbits of F. By the least action principle, lifts of these orbits correspond to critical points of $\mathcal{A}_F$ on $\widetilde{\Lambda M}$. Let $z_\pm \in \mathcal{P}(F)$, and let $\tilde z_\pm = (z_\pm, u_\pm)$ be two lifts of $z_\pm$ to $\widetilde{\Lambda M}$. We denote by $\mathcal{M}(\tilde z_-, \tilde z_+)$ the space of all solutions of (11.5) satisfying the boundary condition

$$u(\tau,t) \xrightarrow[\tau\to\pm\infty]{} z_\pm(t), \tag{11.8}$$

as well as the topological condition

$$u_- \# u \# (-u_+) = 0 \in \overline{\pi}_2(M). \tag{11.9}$$

We call such maps *Floer trajectories* connecting $\tilde{z}_-$ and $\tilde{z}_+$. It turns out that with these boundary conditions, the Floer equation is well posed, and moreover, generically, the spaces of solutions are finite-dimensional manifolds. Since (11.5) is autonomous, $\mathcal{M}(\tilde{z}_-, \tilde{z}_+)$ admits a free $\mathbb{R}$ action by translation. Therefore, one can define the moduli spaces $\overline{\mathcal{M}}(\tilde{z}_-, \tilde{z}_+) = \mathcal{M}(\tilde{z}_-, \tilde{z}_+)/\mathbb{R}$.

11.4. The dimension of moduli spaces

In the finite-dimensional setting, the dimension of the moduli space of connecting trajectories is given by the difference of indices of the critical points. In the infinite-dimensional setting, these indices are in general infinite. However, it turns out that one can associate different indices to 1-periodic orbits such that the above relation still holds.

Recall that in Section 3.4 we defined the Conley–Zehnder index $\mathrm{CZ}(\varphi)$ of a path of symplectic matrices $\varphi\colon [0,1] \to \mathrm{Sp}(2n)$ such that $\varphi(0) = \mathbb{1}$ and $\varphi(1)$ does not have 1 as an eigenvalue. According to the normalization set in Remark 3.4.20, it satisfies the following two properties:

(1) If $H\colon \mathbb{C}^n \to \mathbb{R}$ is a small quadratic form, and h_t its Hamiltonian flow, then $\mathrm{CZ}(h_{t*}) = \operatorname{ind} H$.

(2) If γ is a loop, then

$$\mathrm{CZ}(\varphi \# \gamma) = \mathrm{CZ}(\varphi) - \mathrm{Maslov}(\gamma),$$

where $\mathrm{Maslov}(\gamma)$ is the Maslov index of γ (defined in Remark 3.4.15).

Exercise 11.4.1. Let $\alpha \notin \mathbb{Z}$, identify $\mathbb{C} = \mathbb{R}^2$, and consider the path $\gamma(t) = \mathrm{e}^{2\pi i t\alpha} \in \mathrm{Sp}(2)$ for $t \in [0,1]$. Prove that $\mathrm{CZ}(\gamma) = -2[\alpha]$.

We now define the Conley–Zehnder index for Hamiltonian loops. Let $\tilde{z} = (z, u)$ be a lift of a contractible periodic orbit $z \in \mathcal{P}(F)$. The bundle u^*TM has a contractible base, and hence is trivial as a symplectic vector bundle. Choose a symplectic trivialization

$$u^*TM \simeq D^2 \times (\mathbb{R}^{2n}, \omega_0).$$

Let f_t denote the Hamiltonian flow of F. The trivialization gives an identification of the linearized flow f_{t*} at $z(0)$ with symplectic matrices, i.e. $f_{t*}\big(z(0)\big) \in \mathrm{Sp}(2n)$, and we define

$$\mathrm{CZ}(\tilde{z}) = \mathrm{CZ}\Big(\big\{f_{t*}\big(z(0)\big)\big\}_{t\in[0,1]}\Big).$$

The following relation between the Conley–Zehnder index and the first Chern class will be important in what follows (see [107, Section 2.6]) for the relation between the Maslov index and the first Chern class): for every $A \in \pi_2(M)$ and $(z, u) \in \widetilde{\mathcal{P}}(F)$ one has

$$\mathrm{CZ}(z, u \# A) = \mathrm{CZ}(z, u) + 2c_1(A). \tag{11.10}$$

Theorem 11.4.2. *Assume that $z_\pm \in \mathcal{P}(F)$ are nondegenerate, in the sense that the linearized flow along $z_\pm$ at $t = 1$ does not have 1 as an eigenvalue. Then for a generic pair $\big(F, J(\cdot)\big)$, $\overline{\mathcal{M}}(\tilde z_-, \tilde z_+)$ is a smooth manifold of dimension* $\mathrm{CZ}(\tilde z_-) - \mathrm{CZ}(\tilde z_+) - 1$.

Let us illustrate this formula in a simple local model. Suppose that $(M, \omega) = (\mathbb{C}, \omega_0)$, and take the constant loop $J(t) \equiv \mathrm{i}$. Consider a *time-independent* Hamiltonian $F\colon \mathbb{C} \to \mathbb{R}$ with a pair of nondegenerate constant periodic orbits $z_\pm \in \mathcal{P}(F)$. Then (11.5) and (11.8) become

$$\partial_\tau u + \mathrm{i}\partial_t u + \frac{\partial F}{\partial x}(u) = 0, \quad u(\tau, t) \xrightarrow[\tau\to\pm\infty]{} z_\pm. \tag{11.11}$$

In contrast to the situation described in Remark 11.3.1, this problem turns out to be well posed.

Now, in this special case we may take a t-independent solution $u(\tau)$ to (11.11), which is simply a solution to the negative gradient flow equation

$$\partial_\tau u + \frac{\partial F}{\partial x}(u) = 0 \tag{11.12}$$

with boundary conditions

$$u(\tau) \xrightarrow[\tau\to\infty]{} z_\pm. \tag{11.13}$$

We wish to compute the dimension of the moduli space $\mathcal{M}(z_-, z_+)$ near u. This is the same as the dimension of the tangent space $T_u\mathcal{M}(z_-, z_+)$. So, let $u(\lambda)$ be a path in $\mathcal{M}(z_-, z_+)$ such that $u(0) = u$ and let

$$v = \frac{\mathrm{d}}{\mathrm{d}\lambda}\bigg|_{\lambda=0} u(\lambda) \in T_u\mathcal{M}(z_-, z_+).$$

Differentiating (11.11) with respect to λ gives

$$\partial_\tau v + \mathrm{i}\partial_t v + \frac{\partial^2 F}{\partial x^2}(u)\cdot v = 0, \quad v(\tau, t) \xrightarrow[\tau\to\pm\infty]{} 0. \tag{11.14}$$

Denote

$$\alpha(\tau) = \frac{\partial^2 F}{\partial x^2}\big(u(\tau)\big), \quad \alpha_\pm = \lim_{\tau\to\pm\infty} \alpha(\tau) = \frac{\partial^2 F}{\partial x^2}(z_\pm).$$

Then equation (11.14) becomes

$$\partial_\tau v + \mathrm{i}\,\partial_t v + \alpha(\tau) v = 0, \quad v(\tau, t) \xrightarrow[\tau\to\pm\infty]{} 0.$$

Expand v in a Fourier series:

$$v(\tau, t) = \sum_{n\in\mathbb{Z}} a_n(\tau)\mathrm{e}^{\mathrm{i}nt}, \quad a_n(\tau) \in \mathbb{C},\ t \in [0, 2\pi]. \tag{11.15}$$

Substituting (11.15) into (11.14) yields

$$a_n'(\tau) = a_n(\tau)\big(n - \alpha(\tau)\big).$$

Therefore, if $a_n \not\equiv 0$, we get

$$\frac{a_n'(\tau)}{a_n(\tau)} \xrightarrow[\tau \to \pm\infty]{} n - \alpha_\pm. \tag{11.16}$$

Now, our boundary condition implies that $a_n(\tau) \to 0$ as $\tau \to \pm\infty$. But (11.16) shows that a solution $a_n \not\equiv 0$ with $n > \alpha_+$ will diverge as $\tau \to \infty$, and similarly, if $n < \alpha_-$ a nonzero solution will diverge as $\tau \to -\infty$. So we can have $a_n \not\equiv 0$ only for $\alpha_- < n < \alpha_+$. One readily checks that in this case $a_n(\tau)$ converges to 0 as $\tau \to \pm\infty$. Therefore,

$$\dim_{\mathbb{C}} T_u\mathcal{M}(z_-, z_+) = [\alpha_+] - [\alpha_-]. \tag{11.17}$$

Next, let us compare this with the Conley–Zehnder index difference. Consider the linearized flow at $z_\pm$. The Hamiltonian flow of F is given by the equation

$$\dot{z}(t) = \mathrm{i}\frac{\partial F}{\partial x}\big(z(t)\big).$$

Linearizing at $z_\pm$ gives

$$\dot{w}_\pm(t) = \mathrm{i}\frac{\partial^2 F}{\partial x^2}(z_\pm) \cdot w_\pm(t) = \mathrm{i}\alpha_\pm w_\pm(t). \tag{11.18}$$

Denoting by $\Psi_\pm(t)$ the fundamental solution matrices of (11.18), we have $\mathrm{CZ}(z_\pm) = \mathrm{CZ}(\{\Psi_\pm(t)\})$. Note that the nondegeneracy of $z_\pm$ translates to $\alpha_\pm \notin \mathbb{Z}$. Then by Exercise 11.4.1, $\mathrm{CZ}(z_\pm) = -2[\alpha_\pm]$. Comparing this with (11.17) we get

$$\dim_{\mathbb{R}} T_u\mathcal{M}(z_-, z_+) = 2\big([\alpha_+] - [\alpha_-]\big) = \mathrm{CZ}(z_-) - \mathrm{CZ}(z_+).$$

11.5. Compactness breaking mechanism

In order to define the Floer complex and the Floer differential, we need to understand the compactification of $\overline{\mathcal{M}}(\tilde{x}, \tilde{y})$. As in the case of Morse negative gradient trajectories, one compactness breaking mechanism is convergence of Floer trajectories to broken trajectories, i.e., unions of a finite number of connecting trajectories between intermediate 1-periodic orbits. However, here we have an additional possible failure of compactness, through the *bubbling* phenomenon. A fundamental compactness theorem states that a sequence of Floer trajectories u_k with a uniform C^1-bound on compact sets has a subsequence which converges locally uniformly with all derivatives to a (possibly broken) Floer trajectory. Bubbling occurs when a sequence of connecting trajectories has divergent derivatives: suppose u_k is a sequence of Floer trajectories, and z_k is a sequence of points with $|\mathrm{d}u_k(z_k)| \to \infty$ and $z_k \to z_\infty$. Then one can show that on small neighborhoods of z_∞ suitable reparameterizations of u_k converge to a J-holomorphic map $\mathbb{C} \to M$, and moreover, this map extends holomorphically to ∞. This yields a J-holomorphic sphere, i.e., a J-holomorphic map $S^2 = \mathbb{C} \cup \{\infty\} \to M$, which is called a *bubble*. It turns out that bubbling can happen only at finitely many points, and moreover, away from these points the original sequence

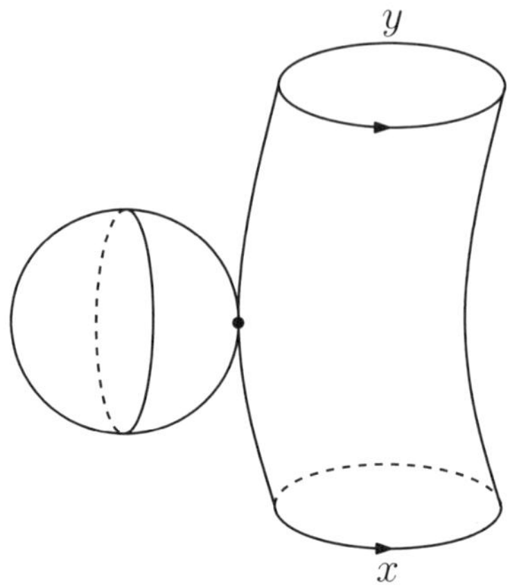

FIGURE 11.2. Bubbling

has a uniform C^1-bound on compacts, and hence converges locally uniformly to a possibly broken Floer trajectory. Let us point out that near each point where the derivatives diverge it is possible for more than one sphere to bubble off, and the resulting bubbles are all connected (this is illustrated in Figure 11.2 for the case of one bubble).

This notion is summarized in the concept of a *cusp trajectory*. This is a tuple $(u, v_1, \ldots, v_l)$, where u is a (possibly broken) Floer connecting trajectory and the v_j are J-holomorphic spheres, such that the images of u and all the v_j form a connected subset of M. For brevity, let us agree to call a possibly broken trajectory simply a broken trajectory. Note that a broken trajectory is a special case of a cusp trajectory with $l = 0$. The Gromov–Floer compactness theorem states that for any sequence of Floer connecting trajectories $u_k \in \mathcal{M}(\tilde{x}, \tilde{y})$ there exists a cusp trajectory $(u_\infty, \{v_j\})$ such that u_k converges to $(u_\infty, \{v_j\})$ along a subsequence. The precise definition of this convergence is rather complicated and we will not discuss it in detail. The relevant part for us is the behaviour of the Conley–Zehnder index under this limit: For an unbroken connecting trajectory u we define the *index* $\Delta_{\mathrm{CZ}}(u)$ to be the Conley–Zehnder index difference between its boundary loops. The index of a broken trajectory u is defined to be the sum of the indices of its components. Then if a sequence $u_k \in \mathcal{M}(\tilde{x}, \tilde{y})$ converge to a cusp trajectory $(u_\infty, v_1, \ldots, v_l)$, for large k the map u_k is homotopic to the connected sum of u_∞ and the spheres $v_1, \ldots, v_l$, and so by the identity (11.10) we have

$$\Delta_{\mathrm{CZ}}(u_\infty) + \sum_{j=1}^{l} 2(c_1, v_j) = \Delta_{\mathrm{CZ}}(u_k) = \mathrm{CZ}(\tilde{x}) - \mathrm{CZ}(\tilde{y}).$$

But, by Exercise 1.6.4, $([\omega], v_j) > 0$ for any J-holomorphic sphere. Therefore the monotonicity of M implies that[2] $(c_1, v_j) > 0$. Since (c_1, v_j) is also an integer, it follows that $(c_1, v_j) \geq 1$. But now if $\mathrm{CZ}(\tilde{x}) - \mathrm{CZ}(\tilde{y}) = 2$, we see

[2]This is a crucial point where one uses the monotonicity assumption on M.

that if $k \geq 1$ the limit connecting trajectory will have $\Delta_{\mathrm{CZ}}(u_\infty) = 0$, which cannot happen generically. Therefore, generically $l = 0$, and so all limit cusp trajectories are simply broken trajectories. Therefore, as in the finite-dimensional case, $\overline{\mathcal{M}}(\tilde{x}, \tilde{y})$ admits a compactification by broken trajectories, which we denote by $\widetilde{\mathcal{M}}(\tilde{x}, \tilde{y})$.

11.6. The Floer complex

Consider the field of Laurent series $\mathcal{K} = \mathbb{Z}_2[[s]$ defined in Section 10.5 above. For a Hamiltonian F consider the lift $\widetilde{\mathcal{P}}(F)$ of the set of all contractible 1-periodic orbits of F to $\widetilde{\Lambda M}$. Define the $\mathcal{K}$-vector space $\mathrm{CF}(F)$ to be the set of all finite linear combination of elements from $\widetilde{\mathcal{P}}(F)$ with coefficients in $\mathcal{K}$, where we identify $s \cdot (z, u)$ with $S \cdot (z, u)$ (defined by (11.1)). Define the differential $\mathrm{d} = \mathrm{d}_J \colon \mathrm{CF}(F) \to \mathrm{CF}(F)$ by counting nonparameterized Floer trajectories:

$$\mathrm{d}\tilde{x} = \sum_{\tilde{y}} n(\tilde{x}, \tilde{y})\tilde{y},$$

where $n(\tilde{x}, \tilde{y})$ vanishes if $\overline{\mathcal{M}}(\tilde{x}, \tilde{y})$ has dimension > 0 and

$$n(\tilde{x}, \tilde{y}) = \#_2 \overline{\mathcal{M}}(\tilde{x}, \tilde{y}) \in \mathbb{Z}_2$$

if $\dim \overline{\mathcal{M}}(\tilde{x}, \tilde{y}) = 0$. In order to check that the differential is well defined, we need the following fact:

Exercise 11.6.1. Prove that the action decreases along Floer trajectories. That is, if $u \in \mathcal{M}(\tilde{z}_-, \tilde{z}_+)$ is a Floer trajectory connecting the lifts $\tilde{z}_\pm = (z_\pm, u_\pm)$ of the periodic orbits $z_\pm$, prove that

$$\mathcal{A}_F(z_-, u_-) \geq \mathcal{A}_F(z_+, u_+).$$

Hint: Compute $\int_{\mathbb{R}\times S^1} u^*\omega$ and use this to show that

$$\mathcal{A}_F(z_-, u_-) - \mathcal{A}_F(z_+, u_+) = \iint_{\mathbb{R}\times S^1} \|\partial_\tau u\|^2_{J(t)}\, \mathrm{d}\tau\, \mathrm{d}t,$$

where $\|\cdot\|_{J(t)}$ denotes the norm determined by the Riemannian metric induced by $J(t)$. Recall the condition (11.9).

Now Exercises 11.2.1 and 11.6.1 ensure, as in the case of Morse–Novikov homology, that $\mathrm{d}\tilde{x}$ is a well-defined element of $\mathrm{CF}(F)$. Furthermore, the discussion on the compactness breaking mechanism presented in Section 11.5 implies that $n(\tilde{x}, \tilde{y})$ is well defined and that $\mathrm{d}^2 = 0$, exactly as in the case of Morse homology.

The matrix coefficient $n(\tilde{x}, \tilde{y}) \neq 0$ only if $\mathrm{CZ}(y) - \mathrm{CZ}(x) = 1$. Thus it is tempting to grade $\mathrm{CF}(F)$ by the Conley–Zehnder index. The caveat is that such a grading is not compatible with the structure of vector space over $\mathcal{K}$. Indeed, by (11.10) $\mathrm{CZ}(s \cdot \tilde{x}) = \mathrm{CZ}(\tilde{x}) + 2N$, where N is the minimal Chern number of M. We shall avoid this difficulty by defining a $\mathbb{Z}_2$-grading. Put $\mathrm{CF}_i(F)$, $i = 0, 1$ to be the subspace of $\mathrm{CF}(F)$ generated by elements with $\mathrm{CZ} = i \mod 2$. We see that $\mathrm{CF}_i(F)$ is indeed a $\mathcal{K}$-subspace, and the differential d shifts the grading by 1 (modulo 2). In this way we get a $\mathbb{Z}_2$-graded chain complex over $\mathcal{K}$. Its homology $H_*(\mathrm{CF}, \mathrm{d})$ is called the *Floer homology* of $\big(F, J(\cdot)\big)$. The construction depends on the generic choice consisting of a 1-periodic Hamiltonian $F\colon M \times S^1 \to M$ and a loop $J(t)$ of ω-compatible almost complex structures. For the sake of brevity, we denote the Floer homology of $\big(F, J(\cdot)\big)$ by $\mathrm{HF}_*(F)$, ignoring the dependence on $J(\cdot)$. As in the finite-dimensional case, all these homologies are pairwise canonically isomorphic as vector spaces over $\mathcal{K}$ and give rise to the Floer homology of M, $\mathrm{HF}_*(M)$.

Remark 11.6.2. Generic Hamiltonians F we deal with satisfy the following condition: the graph of the time-one map $f = f_1$ of the Hamiltonian flow generated by F is transversal to the diagonal in $M \times M$. Therefore, at each fixed point z of F the differential $f_{*z}\colon T_zM \to T_zM$ does not have 1 as an eigenvalue. Exercise 3.4.21 and equation (11.10) above imply that the $\mathbb{Z}_2$-degree of any lift $\tilde{x}$ equals 0 if $\det(f_{*z} - \mathbb{1}) > 0$, and equals 1 otherwise. Thus, strictly speaking, this degree can be defined without the Conley–Zehnder index.

Furthermore, let $\big(F, J(\cdot)\big)$ be a generic pair such that F is a C^∞-small autonomous Morse function, and $J(t) \equiv J$ is a constant loop. In this case, the 1-periodic orbits of the Hamiltonian flow of F are only the constant loops, which are in bijection with the critical points of F. Moreover, one can show that the Floer trajectories are negative gradient trajectories of F. Thus the Floer complex reduces to the Morse complex and we have $\mathrm{HF}_*(F) = \mathrm{HM}_*(F) \otimes_{\mathbb{Z}_2} \mathcal{K}$. Therefore:

Theorem 11.6.3. *There are canonical isomorphisms*

$$\mathrm{HF}_*(M) \simeq \mathrm{HM}_*(M) \otimes_{\mathbb{Z}_2} \mathcal{K} \simeq H_*(M; \mathbb{Z}_2) \otimes_{\mathbb{Z}_2} \mathcal{K}.$$

One of the main motivations of Floer in defining Floer homology was to prove Arnold's conjecture, which we can now do in this setting. Recall that for $f\colon M \to M$, we denote by $\mathrm{Fix}(f)$ the set of fixed points of f, and that $z \in \mathrm{Fix}(f)$ is called nondegenerate if f_{*z} does not have 1 as an eigenvalue.

Theorem 11.6.4 ([67])**.** *Let (M, ω) be a closed spherically monotone symplectic manifold. Let $f \in \mathrm{Ham}(M, \omega)$ be a Hamiltonian diffeomorphism with only nondegenerate fixed points. Then*

$$\#\,\mathrm{Fix}(f) \geq \dim_{\mathbb{Z}_2} H_*(M; \mathbb{Z}_2).$$

Proof. Choose a generic pair $\big(F, J(\cdot)\big)$, where F is a 1-periodic Hamiltonian which generates f. Then the 1-periodic orbits of F correspond to fixed points of f. Therefore, by Theorem 11.6.3,

$$\#\,\mathrm{Fix}(f) = \#\,\mathcal{P}(F) \geq \dim_{\mathcal{K}} \mathrm{HF}_*(F) = \dim_{\mathbb{Z}_2} H_*(M; \mathbb{Z}_2). \qquad \square$$

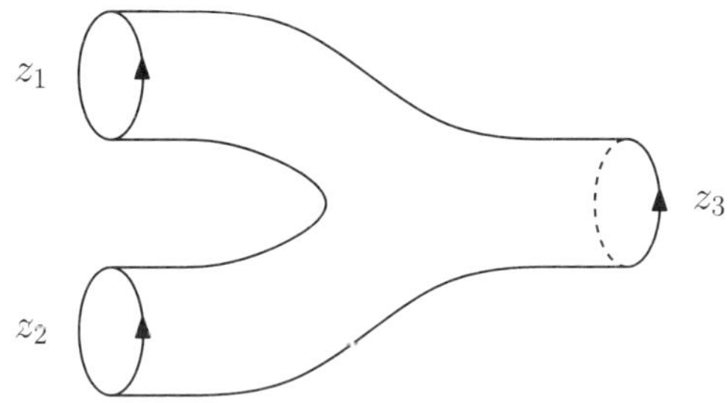

FIGURE 11.3. Pair of pants product

11.7. Ring structure

There is a natural ring structure on Floer homology, which is similar in spirit to the product on Morse homology. Strictly speaking, we shall not use it, but it will be useful to keep it in mind in order to appreciate certain results in the next chapter. Consider *a pair of pants* Σ, which is a noncompact Riemann surface, diffeomorphic to the 2-sphere with three points removed. We think of Σ as the union

$$\Sigma = \Sigma_0 \cup C_1 \cup C_2 \cup C_3,$$

where Σ_0 is diffeomorphic to the 2-sphere with three open discs removed, and the C_i are diffeomorphic to infinite cylinders. Given three generic Hamiltonians F_i, $i = 1, 2, 3$, and 1-periodic orbits $z_i \in \mathcal{P}(F_i)$ with lifts $\tilde{z}_i = (z_i, u_i)$, we consider maps $u \colon \Sigma \to M$ such that:

- On each cylinder C_i, u solves the Floer equation for (F_i, J_i).
- On Σ_0, u satisfies a PDE which, roughly speaking, interpolates between the Floer equations near the boundaries and a Cauchy–Riemann equation in the interior.
- Near the end of each cylinder C_i, u converges to z_i.
- The sphere formed by gluing the discs u_1, u_2, and $-u_3$ to Σ along their boundaries represents the zero class in $\overline{\pi}_2(M)$.

(See Figure 11.3). When the space of such u is zero-dimensional, we define by $n(\tilde{z}_1, \tilde{z}_2, \tilde{z}_3)$ the number, modulo 2, of isolated solutions u of the above problem and we set $n(\tilde{z}_1, \tilde{z}_2, \tilde{z}_3) = 0$ otherwise. We define

$$\mathrm{CF}_*(F_1) \otimes \mathrm{CF}_*(F_2) \to \mathrm{CF}_*(F_3), \quad \tilde{z}_1 \otimes \tilde{z}_2 \mapsto \sum_{\tilde{z}_3} n(\tilde{z}_1, \tilde{z}_2, \tilde{z}_3)\tilde{z}_3.$$

This map induces the pair of pants product

$$\mathrm{HF}_*(F_1) \otimes \mathrm{HF}_*(F_2) \to \mathrm{HF}_*(F_3),$$

which turns out to be associative and commutative.

CHAPTER 12

Constructing Subadditive Spectral Invariants

In this chapter we define the quantum homology ring of a closed symplectic manifold (M,ω), which is a deformation of the singular homology ring of M involving J-holomorphic spheres. The quantum homology turns out to be isomorphic, as a ring, to the Floer homology of the manifold [130]. Using this isomorphism, we describe the construction of *spectral invariants* in Floer theory [119, 144, 157], mimicking the construction of spectral numbers of smooth functions described in Section 10.1. Finally, we use these invariants to prove the existence of subadditive spectral invariants of Hamiltonian diffeomorphisms, asserted in Chapter 4. The exposition in this chapter follows [58, 60].

12.1. Quantum homology

Let (M^{2n},ω) be a closed monotone symplectic manifold. As in previous chapters, we work over the field

$$\mathcal{K} = \mathbb{Z}_2[[s] = \left\{\sum_{i\in\mathbb{Z}} a_i s^i : a_i \in \mathbb{Z}_2,\ a_i = 0 \text{ for all } i > N, \text{ for some } N \in \mathbb{Z}\right\}.$$

As a vector space over $\mathcal{K}$, the quantum homology of M is by definition $\mathrm{QH}(M) = H_*(M;\mathbb{Z}_2)\otimes_{\mathbb{Z}_2}\mathcal{K}$. The field $\mathcal{K}$ contains $\mathbb{Z}_2$ as a subfield. We identify $H_*(M;\mathbb{Z}_2)\otimes_{\mathbb{Z}_2}\mathbb{Z}_2$ with $H_*(M;\mathbb{Z}_2)$ and refer to elements lying in $H_m(M;\mathbb{Z}_2)\subset \mathrm{QH}(M)$ for some $m\in\{0,\dots,2n\}$ as elements of pure degree.

The space QH carries *a quantum product* which is a deformation of the classical intersection product $\cap$ on $H_*(M;\mathbb{Z}_2)$. To understand it, we need to introduce the Gromov–Witten invariants. For $j\in\mathbb{N}$, these are $\mathbb{Z}_2$-trilinear maps

$$\mathrm{GW}_j\colon H_*(M;\mathbb{Z}_2)\times H_*(M;\mathbb{Z}_2)\times H_*(M;\mathbb{Z}_2)\to\mathbb{Z}_2$$

defined as follows. Choose a generic compatible almost complex structure J and generic cycles $\alpha,\beta,\gamma\subset M$ representing classes of pure degree A,B,C, respectively. Consider the space of J-holomorphic curves

$$\mathcal{M}_J(\alpha,\beta,\gamma;j) = \left\{u\colon \mathbb{C}P^1\to M \;\middle|\; \begin{array}{l} u \text{ is } J\text{-holomorphic, } u(0)\in\alpha,\ u(1)\in\beta,\\ u(\infty)\in\gamma,\ [u]=jS\in\overline{\pi}_2(M).\end{array}\right\}$$

(see Figure 12.1). Here j is a nonnegative integer. Indeed, by Exercise 1.6.4, for every J-holomorphic curve u one has $\int_{\mathbb{C}P^1} u^*\omega \geq 0$, hence the space

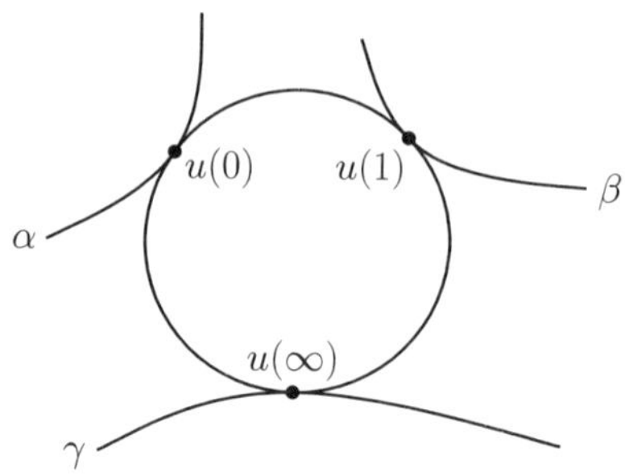

FIGURE 12.1. J-holomorphic spheres passing through α, β, γ

$\mathcal{M}_J(\alpha, \beta, \gamma; j)$ is empty for $j < 0$. Furthermore, every J-holomorphic curve of vanishing symplectic area is constant, and hence

$$\mathcal{M}_J(\alpha, \beta, \gamma; 0) = \alpha \cap \beta \cap \gamma.$$

For a generic J, under the condition

$$\deg A + \deg B + \deg C = 4n - 2Nj \tag{12.1}$$

the space $\mathcal{M}_J(\alpha, \beta, \gamma; j)$ is a compact zero-dimensional manifold. Moreover, in this case its cardinality modulo 2 is independent of the generic choice of J, and we define *the Gromov–Witten invariant* $\mathrm{GW}_j(A, B, C) = \#_2 \mathcal{M}_J(\alpha, \beta, \gamma; j)$. For a ground field other than $\mathbb{Z}_2$, such a count should be performed with appropriate signs. See [108] for further details and proofs.

Example 12.1.1 ($\mathbb{C}P^n$). Let (M, ω, J) be the complex projective space $\mathbb{C}P^n$ equipped with the standard Kähler structure. The generator $S \in \overline{\pi}_2(M)$ is represented by the complex line. The minimal Chern number N equals $n+1$. Denote by $H \in H_{2n-2}(M; \mathbb{Z}_2)$ the class of the hyperplane. Let us compute the Gromov–Witten invariants $\mathrm{GW}_j(H, B, C)$, $j \geq 1$ for classes of pure degree $B, C \in H_*(M; \mathbb{Z}_2)$. By the dimension formula (12.1),

$$2n - 2 + \deg B + \deg C = 4n - (2n+2)j.$$

This leaves us with the only possibility $j = 1$, $\deg B = \deg C = 0$, that is B, C are multiples of the class of the point $P \in H_0(M; \mathbb{Z}_2)$. In order to calculate $\mathrm{GW}_1(H, P, P)$ choose a hyperplane α representing H and a pair of distinct points p_1, p_2 outside α. Denote by ℓ the line passing through p_1 and p_2. The space $\mathcal{M}_J(\alpha, p_1, p_2; 1)$ consists of the unique holomorphic map $u \colon \mathbb{C}P^1 \to \ell$ which sends $0, 1$ and ∞ to $\alpha \cap \ell$, p_1 and p_2, respectively. Therefore $\mathrm{GW}_1(H, P, P) = 1$.

Write $[M]$ for the fundamental class of M.

Proposition 12.1.2. $\mathrm{GW}_j(A, B, [M]) = 0$ *for every classes* $A, B \in H_*(M; \mathbb{Z}_2)$ *and* $j \geq 1$.

PROOF. Without loss of generality, assume that A and B have pure degree. Let α and β be generic cycles representing the classes A and B, respectively. We claim that for every $j > 0$ the space $X := \mathcal{M}_J(\alpha, \beta, M; j)$ cannot be a nonempty zero-dimensional manifold. Indeed, assume on the contrary that X contains a J-holomorphic curve $u\colon \mathbb{C}P^1 \to M$. Denote by $\Theta \subset \mathrm{PSL}(2, \mathbb{C})$ the stabilizer of the points 0 and 1 which is isomorphic to the complex circle $\mathbb{C}_* = \mathbb{C} \setminus \{0\}$. The constraint $u(\infty) \in M$ is void, and hence $u \circ \theta \in X$ for every $\theta \in \Theta$. Observe that since $j > 0$, the curve u is nonconstant. Hence, there exists a point $x \in \mathbb{C}P^1$ such that u restricted to a neighborhood of x is one to one. Without loss of generality $x \notin \{0, 1\}$. Since Θ acts transitively on $\mathbb{C}P^1 \setminus \{0, 1\}$, we conclude that X contains a two-dimensional piece and hence cannot be 0-dimensional, a contradiction. □

For $j \in \mathbb{N}$ and $A, B \in H_*(M; \mathbb{Z}_2)$, we define $(A * B)_j \in H_*(M; \mathbb{Z}_2)$ by the condition

$$(A * B)_j \circ C = \mathrm{GW}_j(A, B, C) \quad \forall C \in H_*(M; \mathbb{Z}_2),$$

where $\circ$ denotes the intersection index modulo 2. The *quantum product* on $\mathrm{QH}(M)$ is given by

$$A * B = A \cap B + \sum_{j \in \mathbb{N}} (A * B)_j s^{-j}.$$

This product is commutative[1] and moreover associative (the latter is a non-trivial statement, see, e.g., [108] for a proof). Proposition 12.1.2 yields that the fundamental class $[M]$ is the unit element with respect to the quantum product. We shall write $[M] = 1$.

Example 12.1.3 ($\mathbb{C}P^n$ continued). Let us calculate the quantum product on the ring $\mathrm{QH}(\mathbb{C}P^n)$. As in Example 12.1.1 above, we write H and P for the classes of hyperplane and point respectively. Since $\mathrm{GW}_1(H, P, P)$ is the only nonvanishing Gromov–Witten invariant involving H, we have that $H * P = (H * P)_1 s^{-1}$, where $(H * P)_1 \circ P = 1$ and $(H * P)_1 \circ C = 0$ for every class C of positive pure degree. It follows that $(H * P)_1 = 1$ and hence $H * P = H^{n+1} = 1 \cdot s^{-1}$. We conclude that as an algebra over $\mathcal{K}$

$$\mathrm{QH}(\mathbb{C}P^n) = \mathcal{K}[H]/\{H^{n+1} = s^{-1}\}.$$

Here $\mathcal{K}[H]$ is the ring of polynomials in the formal variable H with coefficients in $\mathcal{K}$. Since the polynomial $f(H) = H^{n+1} - s^{-1}$ is irreducible, we conclude that $\mathrm{QH}(\mathbb{C}P^n)$ is a field.

Example 12.1.4 ($\mathbb{T}^4 \sharp \overline{\mathbb{C}P^2}$, McDuff [60, 104]). Let (M, J) be the complex blow up of the complex torus $\mathbb{T}^4 = \mathbb{C}^2/\mathbb{Z}^4$ at some point $z \in \mathbb{T}^4$. The manifold M carries a monotone symplectic structure ω compatible with J, see [107] for a discussion on symplectic blow up. The generator S of $\overline{\pi}_2(M)$ is represented by the exceptional divisor $\Sigma \subset M$. Denote by E the homology class of Σ. Note that $c_1(E) = 1$, so that the minimal Chern number N equals 1.

[1]In general it is skew-commutative, but since we are working with $\mathbb{Z}_2$ coefficients, it is commutative.

The homology $H_*(M;\mathbb{Z}_2)$ is spanned by the class E, the fundamental class $[M]=1$, and the space $V := \bigoplus_{i=0}^{3} H_i(\mathbb{T}^4;\mathbb{Z}_2)$. Denote by P the class of point.

We claim that the only nonconstant holomorphic spheres in (M,J) are given by Σ and its multiple covers. Indeed, write $\tau\colon M \to \mathbb{T}^4$ for the natural projection so that $\Sigma = \tau^{-1}(z)$. For a nonconstant holomorphic sphere $C \subset M$, the projection $\tau(C)$ is a holomorphic sphere in $\mathbb{T}^4$. Since $\mathbb{T}^4$ carries no nonconstant holomorphic spheres, the only possibility is that $\tau(C) = z$, which yields the claim.

It follows that all Gromov–Witten invariants except maybe $\mathrm{GW}_j(E,E,E)$ vanish. Thus $E*A = 0$ and $A*B = A\cap B$ for all $A,B\in V$. By the dimension formula (12.1), $\mathrm{GW}_j(E,E,E) = 0$ for $j\geq 2$. Choose three generic cycles $\alpha_1,\alpha_2,\alpha_3$ representing E. In particular, the genericity yields that α_k intersects Σ transversally in m_k points, and these points are distinct for different k's. Since Σ is the unique nonparameterized holomorphic sphere in the class S, the space $\mathcal{M}_J(\alpha,\beta,\gamma;1)$ consists of exactly $m_1m_2m_3$ points. Recall now that $E\circ E=-1$ which yields that each m_k is odd. Thus $\mathrm{GW}_1(E,E,E)=1$, and hence $(E,E)_1 = E$. We conclude that $E*E = p + s^{-1}E$, which completes the quantum multiplication table for $\mathrm{QH}(M)$. Put $e := s^2P+sE$. One can readily check that e is an idempotent. Furthermore, $\mathrm{QH}(M)$ splits as an algebra as $\mathcal{F}\oplus\mathcal{R}$, where $\mathcal{F} := \mathrm{Span}_{\mathbb{Z}_2}(e)$ is a field and $\mathcal{R} = \mathrm{Span}_{\mathbb{Z}_2}(1-e,V)$. Let us emphasize that $\mathrm{QH}(M)$ is not a field (for instance, $P*P=0$), however it possesses a field factor $\mathcal{F}$.

12.2. The Frobenius structure

Recall that a finite-dimensional unital algebra A over a filed $\mathcal{K}$ is called *Frobenius* if it admits a nondegenerate bilinear form $\Delta\colon A\times A\to\mathcal{K}$ which satisfies

$$\Delta(ab,c) = \Delta(a,bc). \tag{12.2}$$

The quantum homology algebra $\mathrm{QH}(M)$ carries a structure of Frobenius algebra. The form Δ is given by

$$\Delta\Big(\sum A_j s^j, \sum B_l s^l\Big) = \sum (A_j\circ B_l)\cdot s^{j+l}.$$

Nondegeneracy of Δ readily follows from that of the intersection index. Let us check the Frobenius condition (12.2). Since both Δ and the quantum product $*$ are linear over $\mathcal{K}$, it suffices to verify (12.2) for a triple of pure degree classes $A,B,C\in H_*(M;\mathbb{Z}_2)$. By definition,

$$\Delta(A*B,C) = \sum_{j\geq 0}((A*B)_j\circ C)\cdot s^{-j} = \sum_{j\geq 0}\mathrm{GW}_j(A,B,C)s^{-j} = \Delta(B*C,A).$$

The equality on the right follows from the fact that the Gromov–Witten invariants are symmetric. Since Δ is symmetric, $\Delta(B*C,A) = \Delta(A,B*C)$, which yields (12.2).

Let us note finally that $\Delta(A, B) = \Delta(A * B, 1) = (A * B) \circ [M]$. Thus $\Delta(A, B)$ can be interpreted as the coefficient at $P = [\text{point}]$ in the quantum product $A * B$.

12.3. Non-Archimedean geometry of QH

Define a function $\sigma\colon \mathcal{K} \to \mathbb{Z}$ as follows. For a nonzero element $a = \sum a_j s^j \in \mathcal{K}$ set $\sigma(a)$ to be the maximal j such that $a_j \neq 0$. Our convention is that $\sigma(0) = -\infty$. Note that $\sigma(a + b) \leq \max\big(\sigma(a), \sigma(b)\big)$ and $\sigma(ab) = \sigma(a) + \sigma(b)$. It follows that $|a| := \exp \sigma(a)$ is an *non-Archimedean absolute value* on $\mathcal{K}$:

(i) $|0| = 0$, $|a| > 0$ for $a \neq 0$;
(ii) $|a + b| \leq \max(|a|, |b|)$;
(iii) $|ab| = |a| \cdot |b|$.

For preliminaries on non-Archimedean geometry the reader is referred to [98, Chapter XII].

Exercise 12.3.1. Prove that $\mathcal{K}$ is complete with respect to this absolute value: every Cauchy sequence has a limit.

The field $\mathcal{K}$ embeds as a subfield into $\mathrm{QH}(M)$, $a \mapsto [M] \otimes a$. The function σ extends to $\mathrm{QH}(M)$ in the following way: For a nonzero element $A = \sum A_j s^j \in \mathrm{QH}(M)$, we define $\sigma(A)$ as the maximal j such that $A_j \neq 0$. The main result of the present section is as follows.

Proposition 12.3.2. *Assume that* $\mathrm{QH}(M)$ *splits as an algebra as* $\mathcal{F} \oplus \mathcal{R}$, *where* $\mathcal{F}$ *is a field. Then there exists* $r \in \mathbb{R}$ *such that* $\sigma(b) + \sigma(b^{-1}) \leq r$ *for every* $b \in \mathcal{F} \setminus \{0\}$.

Proof.[2] The function $\|A\| := \exp \sigma(A)$ is a norm on $\mathrm{QH}(M)$ considered as a vector space over $\mathcal{K}$. Since the field $\mathcal{F}$ is a finite extension of $\mathcal{K}$, the absolute value $|\cdot|$ extends to an absolute value $|||\cdot|||$ on $\mathcal{F}$ (see Proposition 2.5 in [98, Chapter XII]). Furthermore, all norms on a finite-dimensional space over $\mathcal{K}$ are equivalent. Thus there exists $\delta > 0$ so that

$$\|b\| \leq \delta \cdot |||b|||$$

for every $b \in \mathcal{F}$. Therefore for $b \neq 0$

$$\|b\| \cdot \|b^{-1}\| \leq \delta^2 \cdot |||b||| \cdot |||b^{-1}||| = \delta^2,$$

where the last equality follows from the definition of the absolute value. Therefore $\sigma(b) + \sigma(b^{-1}) \leq r$, with $r = 2 \log \delta$. This completes the proof. □

[2]This argument, which is borrowed from [58], is due to V. Berkovich.

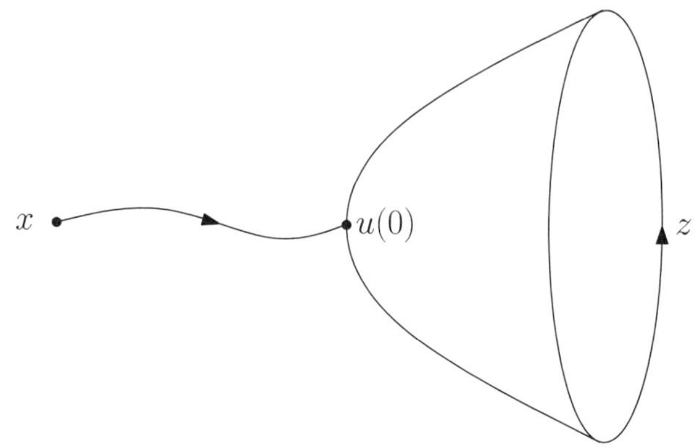

FIGURE 12.2. Spiked discs

12.4. The PSS isomorphism

It turns out that the quantum homology is isomorphic, *as a ring*, to the Floer homology of M. An explicit isomorphism is given by the Piunikhin–Salamon–Schwarz (which we abbreviate to PSS) isomorphism [108, 130]. Let us sketch the construction of this map. First, choose generic data: a Morse function F, a Riemannian metric ρ, a 1-periodic Hamiltonian H, and an almost complex structure J. Now choose a critical point $x \in \mathrm{Crit}(F)$, a 1-periodic orbit $z \in \mathcal{P}(H)$, and a lift $\tilde{z} = (z, v)$ of z.

The PSS map is defined by considering maps $u\colon \mathbb{C} \to M$ which satisfy the following conditions:

- The map u is J-holomorphic near the origin.
- The map $u(se^{2\pi i t})$ satisfies the Floer equation for H for large s, and converges to $z(t)$ as $s \to \infty$.
- The point $u(0)$ converges to x in negative time under the negative gradient flow of F. In other words, $u(0) \in W^{\mathrm{u}}(x)$.
- The sphere $u \# (-v)$ represents the zero class in $\overline{\pi}_2(M)$.

Such maps u are called *spiked discs* (see Figure 12.2). When the space of spiked discs is zero-dimensional, we denote its cardinality by $n(x, \tilde{z})$, and define a map

$$\Phi_{\mathrm{PSS}}\colon \mathrm{CM}_*(F) \otimes \mathcal{K} \to \mathrm{CF}_*(H), \quad x \mapsto \sum_{\tilde{y}} n(x, \tilde{y})\tilde{y}.$$

One can show that Φ_{PSS} is a chain map, which induces an isomorphism of vector spaces

$$\mathrm{QH}_*(M) \simeq \mathrm{HM}_*(F) \otimes \mathcal{K} \to \mathrm{HF}_*(M).$$

A deep fact is that the PSS map is in fact an isomorphism of *rings*.

12.5. Spectral invariants in Floer theory

The spectral invariants in Floer theory are defined by following the same scheme as that of the Morse-theoretic spectral numbers, and satisfy similar properties, although the proofs of these properties are much harder in this case [119, 144, 157].

Consider (M, ω) be a spherically monotone symplectic manifold. Let $(F, J(\cdot))$ be a generic pair consisting of a smooth 1-periodic Hamiltonian on F and a loop of compatible almost complex structures $J(t)$. The Floer chain complex $\mathrm{CF}(F)$ is filtered by the action functional, and Exercise 11.6.1 implies that the Floer differential preserves this filtration. Thus we have a well-defined Floer homology of the sublevel sets $\{\mathcal{A}_F < \alpha\}$, which we denote by $\mathrm{HF}(\alpha)$. The inclusion on chains induces homomorphisms

$$i_\alpha\colon \mathrm{HF}(\alpha) \to \mathrm{HF}(M).$$

Using the PSS isomorphism, we identify $\mathrm{HF}(M) \simeq \mathrm{QH}(M)$, and given $a \in \mathrm{QH}(M)$ we define

$$c(a, F) = \inf\{\alpha : a \in \operatorname{Image} i_\alpha\}.$$

Let us note that although the Floer complex depends on the choice of the loop of complex structures $J(t)$, a continuation argument proves that different choices of such loops yield complexes with naturally isomorphic homology. Moreover, this isomorphism is filtration preserving, and hence the definition of $c(a, F)$ does not depend on $J(t)$.

As noted above, the Floer-theoretic spectral invariants enjoy properties similar to those of their Morse-theoretic counterparts. Before we explore these, we prove the following important fact:

Proposition 12.5.1. *For a normalized Hamiltonian F, the spectral invariant $c(a, F)$ depends only on the element $\phi \in \widetilde{\mathrm{Ham}}(M)$ generated by F.*

Proof. Let $\{h_t\}_{t\in[0,1]}$ be a contractible loop in $\mathrm{Ham}(M, \omega)$, generated by the normalized Hamiltonian $H(x, t)$. We must show that $c(a, F) = c(a, F_h)$, where

$$F_h(x, t) = -H(h_t x, t) + F(h_t x, t)$$

is the normalized Hamiltonian generating the path $\{h_t^{-1} f_t\}$. Define

$$T_h\colon \Lambda M \to \Lambda M, \quad z(t) \mapsto h_t z(t).$$

Since $\{h_t\}$ is contractible, this lifts to a map $\widetilde{T}_h\colon \widetilde{\Lambda M} \to \widetilde{\Lambda M}$.

Exercise 12.5.2. (i) Prove that $(\widetilde{T}_h^{-1})^* \mathcal{A}_F = \mathcal{A}_{F_h}$. *Hint*: The proof is similar to that of Proposition 4.2.1 above.

(ii) Show that T_h establishes a bijection between solutions to the Floer equations of the pairs (F_h, J) and (F, J_h), where $J_h(t) = h_{t*} J(t) (h_{t*})^{-1}$.

Therefore, for a generic loop $J(t)$ of compatible almost complex structures, $\widetilde{T}_h$ induces an isomorphism

$$(\mathrm{CF}(F), \mathrm{d}_{J_h}) \simeq (\mathrm{CF}(F_h), \mathrm{d}_J).$$

Moreover, this isomorphism preserves the action filtration. Therefore, $c(a, F) = c(a, F_h)$. $\square$

In view of Proposition 12.5.1, we use the notation $c(a, \phi) := c(a, F)$, where $\phi \in \widetilde{\mathrm{Ham}}(M)$ is generated by a normalized Hamiltonian F.

Next, we list several important properties of the spectral invariants established by Oh [108, 119, 120] and, in maximal generality, by Usher [154, 155]. Let us mention that these properties are parallel to the ones of spectral numbers of functions on manifolds, see Section 10.1.

Proposition 12.5.3. *The spectral invariants satisfy the following properties:*

(1) **(Spectrality)**[3] *The spectral invariant $c(a, \phi)$ belongs to the action spectrum of ϕ.*
(2) **(Subadditivity)** *For any nonzero classes $a, b \in \mathrm{QH}(M)$ and elements $\varphi, \psi \in \widetilde{\mathrm{Ham}}(M)$ we have*
$$c(a * b, \varphi\psi) \leq c(a, \varphi) + c(b, \psi).$$
(3) **(Stability)** *For any normalized Hamiltonians F_t, G_t generating elements $\phi, \psi \in \widetilde{\mathrm{Ham}}(M)$ and a nonzero class $a \in \mathrm{QH}(M)$,*
$$\int_0^1 \min_M (F_t - G_t)\, \mathrm{d}t \leq c(a, \phi) - c(a, \psi) \leq \int_0^1 \max_M (F_t - G_t)\, \mathrm{d}t.$$
(4) **(Value at the identity)** $c(a, \mathbb{1}) = \Omega \cdot \sigma(a)$ *for every nonzero class $a \in \mathrm{QH}(M)$, where $\Omega = ([\omega], S)$.*
(5) **(Conjugation invariance)** *For any nonzero $a \in \mathrm{QH}(M)$ and any $\phi, \psi \in \widetilde{\mathrm{Ham}}(M)$,*
$$c(a, \phi\psi\phi^{-1}) = c(a, \psi).$$

Finally, the spectral invariants satisfy a form of Poincaré duality [58, 155]. In order to state it, introduce a function $\tau \colon \mathcal{K} \to \mathbb{Z}_2$ as the map sending $\sum z_j s^j$ to z_0. Define a $\mathbb{Z}_2$-valued pairing

$$\Pi(a, b) = \tau\Delta(a, b) = \tau\Delta(a * b, [M]) \tag{12.3}$$

on $\mathrm{QH}(M)$, where Δ is the Frobenius form, see Section 12.2. For an element $b \in \mathrm{QH}(M) \setminus \{0\}$, denote by $\Upsilon(b)$ the set of all $a \in \mathrm{QH}(M)$ with $\Pi(a, b) \neq 0$. With this language we have

(6) **(Poincaré duality)** $c(b, \phi) = -\inf_{a \in \Upsilon(b)} c(a, \phi^{-1})$ *for all $b \in \mathrm{QH}(M) \setminus \{0\}$ and $\phi \in \widetilde{\mathrm{Ham}}(M)$.*

12.6. Subadditive spectral invariants revisited

Having set up the theory of Floer-theoretic spectral invariants, we can now turn our attention back to the question of existence of subadditive spectral invariants on $\widetilde{\mathrm{Ham}}(M, \omega)$ and their properties. Our first observation

[3]Spherically monotone symplectic manifolds enjoy an "upgraded" version of the spectrality, see Remark 4.3.2.

is that if $e \in \mathrm{QH}(M) \setminus \{0\}$ is a nonvanishing idempotent, that is $e^2 = e$, the function

$$c\colon \widetilde{\mathrm{Ham}}(M) \to \mathbb{R}, \quad \phi \mapsto c(e, \phi)$$

is a subadditive spectral invariant. Indeed, the conjugation invariance, stability and spectrality axioms are immediate from Proposition 12.5.3. The subadditivity axiom follows from the inequality:

$$c(e, \phi\psi) = c(e * e, \phi\psi) \leq c(e, \phi) + c(e, \psi).$$

We can now prove Theorem 4.7.1, which asserts the existence of a subadditive spectral invariant with $c(\mathbb{1}) = 0$, for monotone closed symplectic manifolds. Indeed, taking the idempotent $e = [M] = 1$ we have $c(e, \mathbb{1}) = 0$, and so the function $\phi \mapsto c(e, \phi)$ is the desired subadditive spectral invariant.

The next result gives a sufficient condition for existence of a subadditive spectral invariant with bounded spectral pseudo-norm.

Theorem 12.6.1. *Let (M, ω) be a closed monotone symplectic manifold. Assume that the quantum cohomology $\mathrm{QH}(M)$ has a field factor, that is, it splits as an algebra as $\mathcal{F} \oplus \mathcal{R}$, where $\mathcal{F}$ is a field. Let $c(\phi) := c(e, \phi)$ be the subadditive spectral invariant associated to the unity e of $\mathcal{F}$. Then the corresponding spectral pseudo-norm ν_c is bounded.*

Proof. Recall that the spectral pseudo-norm ν_c is defined by $\nu_c(\phi) = c(\phi) + c(\phi^{-1})$ for all $\phi \neq \mathbb{1}$ and $\nu_c(\mathbb{1}) = 0$. Let us find a lower bound for $c(b, \phi)$ provided $\Pi(b, e) \neq 0$. Write $b = b_1 + b_2$, where $b_1 \in \mathcal{F}$ and $b_2 \in \mathcal{R}$. Note that $b * e = b_1 * e = b_1$, and

$$\Pi(b, e) = \tau\Delta(b * e, [M]) = \tau\Delta(b_1, [M]) \neq 0.$$

This implies that $b_1 \neq 0$ and $\sigma(b_1) \geq 0$. Therefore, b_1 is invertible in $\mathcal{F}$ and $\sigma(b_1^{-1}) \leq r$, where r is the constant from Lemma 12.3.2. Thus, by the subadditivity,

$$\begin{aligned} c(b, \phi) &\geq c(b * e, \phi) - c(e, \mathbb{1}) = c(b_1 * e, \phi) - c(e, \mathbb{1}) \\ &\geq c(e, \phi) - c(b_1^{-1}, \mathbb{1}) - c(e, \mathbb{1}). \end{aligned}$$

But

$$c(b_1^{-1}, \mathbb{1}) = \Omega \cdot \sigma(b_1^{-1}) \leq \Omega r \quad \text{and} \quad c(e, \mathbb{1}) = \Omega\sigma(e).$$

Therefore,

$$c(b, \phi) \geq c(e, \phi) - \Omega r - \Omega\sigma(e).$$

By Poincaré duality,

$$c(e, \phi^{-1}) = - \inf_{b:\Pi(b,e)\neq 0} c(b, \phi) \leq -c(e, \phi) + \Omega r + \Omega\sigma(e).$$

Therefore,

$$c(e, \phi^{-1}) + c(e, \phi) \leq \Omega r + \Omega\sigma(e)$$

for all $\phi \in \widetilde{\mathrm{Ham}}(M)$, as required. □

As an immediate corollary, from Theorem 4.8.1 we get the existence of a Calabi quasi-morphism in the conditions of Theorem 12.6.1.

Corollary 12.6.2. *Let (M, ω) be a closed monotone symplectic manifold, and assume that its quantum homology* $\mathrm{QH}(M)$ *has a field factor. Then there exists a Calabi quasi-morphism on* $\widetilde{\mathrm{Ham}}(M, \omega)$.

Example 12.6.3. Consider $\mathbb{C}P^n$ with the Fubini-Study form. By Example 12.1.1, the quantum homology $\mathrm{QH}(\mathbb{C}P^n)$ is a field with unit $e = [\mathbb{C}P^n]$. We conclude that the subadditive spectral invariant $c(\phi) = c(e, \phi)$ has bounded spectral pseudo-norm, and hence its normalized homogenization

$$\mu(\phi) = -\operatorname{Vol}(\mathbb{C}P^n) \cdot \lim_{k \to \infty} \frac{c(\phi^k)}{k}$$

is a Calabi quasi-morphism.

Now we can turn to the special case of $\mathbb{C}P^1 = S^2$, and complete the proof of Proposition 5.1.10:

PROOF OF PROPOSITION 5.1.10. We must show that the median quasi-state on S^2 is the quasi-state associated to a subadditive spectral invariant with bounded spectral pseudo-norm. Indeed, by Example 12.6.3, the subadditive spectral invariant $c(\phi) := c(e, \phi)$ (where $e = [S^2]$) has bounded spectral pseudo-norm. Thus, by Proposition 5.1.2, c generates a (genuine!) symplectic quasi-state, say η. Moreover, η has the vanishing property (see Section 4.5). But by Exercise 5.4.29 the median quasi-state ζ is the unique symplectic quasi-state on S^2 with the vanishing property. Therefore $\eta = \zeta$, which completes the proof. □

We conclude this chapter with a remark concerning the choice of coefficients in the Floer and quantum homologies. For the sake of simplicity of the exposition we dealt all the way with $\mathbb{Z}_2$-coefficients. In a number of interesting examples this simplification leads to a significant loss of information. In order to explain this, let us start with quantum homology of the two-sphere $S^2 = \mathbb{C}P^1$. It is generated by the class of the point P (which is the same as the class of the hyperplane) and by the fundamental class, denoted by 1. The relation between the generators is $P^2 = s^{-1} \cdot 1$. The polynomial $P^2 - s^{-1} \cdot 1$ in the variable P is irreducible over either of the fields $\mathbb{Z}_2[[s]$ and $\mathbb{C}[[s]$ of semi-infinite Laurent series with $\mathbb{Z}_2$- and $\mathbb{C}$-coefficients, respectively. Thus $\mathrm{QH}(S^2)$ is a field independently of the choice of a ground field.

Turn now to the case of the product $M = S^2 \times S^2$, where the symplectic form is split and the areas of the factors are equal, which guarantees monotonicity. By a version of the Künneth formula for quantum homology (see [108]), $\mathrm{QH}(M)$ is generated by the four classes

$$p := P \otimes P, \quad a := P \otimes 1, \quad b := 1 \otimes P, \quad e := 1 \otimes 1,$$

and the quantum product is defined by

$$(A_1 \otimes B_1) * (A_2 \otimes B_2) = (A_1 * A_2) \otimes (B_1 * B_2).$$

Note that e is the unit element of $\mathrm{QH}(M)$. We claim that $\mathrm{QH}(M)$ has no field factors when defined over $\mathbb{Z}_2[[s]$ and that it splits into the sum of two fields over $\mathbb{C}[[s]$.

In order to check the first claim, note that $a^2 = b^2 = s^{-1}e$ and $p^2 = s^{-2}e$. Write an arbitrary element $u \in \mathrm{QH}(M)$ in the form

$$u = \alpha \cdot e + \beta \cdot a + \gamma \cdot b + \delta \cdot p, \quad \alpha, \beta, \gamma, \delta \in \mathbb{Z}_2[[s],$$

and look at the equation $u^2 = u$ defining an idempotent. We calculate that over $\mathbb{Z}_2[[s]$ it is equivalent to

$$(\alpha^2 + (\beta^2 + \gamma^2)s^{-1} + \delta^2 s^{-2})e = \alpha \cdot e + \beta \cdot a + \gamma \cdot b + \delta \cdot p,$$

which immediately yields $\beta = \gamma = \delta = 0$. It follows that e is the only nonvanishing idempotent of $\mathrm{QH}(M)$ over $\mathbb{Z}_2[[s]$. At the same time $(p + s^{-1}e)^2 = 0$ and hence $\mathrm{QH}(M)$ is not a field. We conclude that $\mathrm{QH}(M)$ over $\mathbb{Z}_2[[s]$ has no field factors, and the claim follows.

Let us turn to the second claim.

Exercise 12.6.4. Prove that over $\mathbb{C}[[s]$ the quantum homology $\mathrm{QH}(M)$ decomposes (as an algebra) into the direct sum of two fields with unit elements

$$e_\pm = \frac{e \pm ps}{2}.$$

The quasi-measure τ_- corresponding to the idempotent e_- distinguishes the Clifford torus in M from the exotic Lagrangian torus K which was discussed in Example 6.2.4 above.

Let us mention finally that in order to work over $\mathbb{C}[[s]$, one has to choose *coherent* orientations on various moduli spaces arising in the Floer theory and in the quantum homology. One can do this following the approach developed by Floer and Hofer in [68].

Bibliography

1. J. F. Aarnes, *Quasi-states and quasi-measures*, Adv. Math. **86** (1991), no. 1, 41–67.
2. M. Akveld and D. Salamon, *Loops of Lagrangian submanifolds and pseudoholomorphic discs*, Geom. Funct. Anal. **11** (2001), no. 4, 609–650.
3. S. T. Ali, C. Carmeli, T. Heinosaari, and A. Toigo, *Commutative POVMs and fuzzy observables*, Found. Phys. **39** (2009), no. 6, 593–612.
4. V. I. Arnold, *Sur une propriété topologique des applications globalement canoniques de la mécanique classique*, C. R. Acad. Sci. Paris **261** (1965), 3719–3722.
5. ______, *On a characteristic class entering into conditions of quantization* (Russian), Funkcional. Anal. i Prilozen. **1** (1967), 1–14.
6. ______, *A stability problem and ergodic properties of classical dynamical systems* (Russian), Proc. Internat. Congr. Math. (Moscow, 1966), Izdat. "Mir", Moscow, 1968, pp. 387–392.
7. ______, *Mathematical methods of classical mechanics*, translated by K. Vogtmann and A. Weinstein, Grad. Texts in Math., vol. 60, Springer, New York, 1978.
8. ______, *The first steps of symplectic topology* (Russian), Uspekhi Mat. Nauk **41** (1986), no. 6(252), 3–18, 229; English transl., Russian Math. Surveys **41** (1986), no. 6, 1–21.
9. E. Arthurs and M. S. Goodman, *Quantum correlations: a generalized Heisenberg uncertainty relation*, Phys. Rev. Lett. **60** (1988), no. 24, 2447–2449.
10. E. Arthurs and J. L. Kelly Jr., *On the simultaneous measurement of a pair of conjugate observables*, Bell System Tech. J. **44** (1965), 725–729.
11. M. F. Atiyah, *Convexity and commuting Hamiltonians*, Bull. London Math. Soc. **14** (1982), no. 1, 1–15.
12. M. Audin and M. Damian, *Théorie de Morse et homologie de Floer*, Savoirs Actuels (Les Ulis), EDP Sciences, Les Ulis; CNRS Éditions, Paris, 2010.
13. A. Banyaga, *Sur la structure du groupe des difféomorphismes qui préservent une forme symplectique*, Comment. Math. Helv. **53** (1978), no. 2, 174–227.
14. A. Banyaga and D. Hurtubise, *Lectures on Morse homology*, Kluwer Texts Math. Sci., vol. 29, Kluwer Acad. Publ., Dordrecht, 2004.
15. J. Barge and É. Ghys, *Surfaces et cohomologie bornée*, Invent. Math. **92** (1988), no. 3, 509–526.
16. A. Barvinok, *A course in convexity*, Grad. Stud. Math., vol. 54, Amer. Math. Soc., Providence, RI, 2002.
17. C. Bavard, *Longueur stable des commutateurs*, Enseign. Math. (2) **37** (1991), no. 1-2, 109–150.
18. J. S. Bell, *On the problem of hidden variables in quantum mechanics*, Rev. Modern Phys. **38** (1966), 447–452.
19. G. Ben Simon, *The nonlinear Maslov index and the Calabi homomorphism*, Commun. Contemp. Math. **9** (2007), no. 6, 769–780.
20. S. K. Berberian, *Notes on spectral theory*, Van Nostrand Mathematical Studies, vol. 5, Van Nostrand, Princeton, NJ–Toronto, ON–London, 1966.

21. F. A. Berezin, *General concept of quantization*, Comm. Math. Phys. **40** (1975), 153–174.
22. P. Biran, *A stability property of symplectic packing*, Invent. Math. **136** (1999), no. 1, 123–155.
23. ———, *From symplectic packing to algebraic geometry and back*, European Congress of Mathematics, Vol. II (Barcelona, 2000) (C. Casacuberta, R. M. Miró-Roig, J. Verdera, and S. Xambó-Descamps, eds.), Progr. Math., vol. 202, Birkhäuser, Basel, 2001, pp. 507–524.
24. P. Biran, M. Entov, and L. Polterovich, *Calabi quasimorphisms for the symplectic ball*, Commun. Contemp. Math. **6** (2004), no. 5, 793–802.
25. M. Bordemann, E. Meinrenken, and M. Schlichenmaier, *Toeplitz quantization of Kähler manifolds and* $\mathrm{gl}(N)$, $N \to \infty$ *limits*, Comm. Math. Phys. **165** (1994), no. 2, 281–296.
26. M. S. Borman, *Symplectic reduction of quasi-morphisms and quasi-states*, J. Symplectic Geom. **10** (2012), no. 2, 225–246.
27. ———, *Quasi-states, quasi-morphisms, and the moment map*, Int. Math. Res. Not. IMRN **11** (2013), 2497–2533.
28. M. S. Borman and F. Zapolsky, *Quasi-morphisms on contactomorphism groups and contact rigidity*, Geom. Topol., to appear, available at `arXiv:1308.3224`.
29. D. Borthwick and A. Uribe, *Almost complex structures and geometric quantization*, Math. Res. Lett. **3** (1996), no. 6, 845–861.
30. R. Bott and L. W. Tu, *Differential forms in algebraic topology*, Grad. Texts in Math., vol. 82, Springer, New York, 1982.
31. M. Brandenbursky, *Bi-invariant metrics and quasi-morphisms on groups of Hamiltonian diffeomorphisms of surfaces*, available at `arXiv:1306.3350`.
32. L. Buhovsky, *The 2/3-convergence rate for the Poisson bracket*, Geom. Funct. Anal. **19** (2010), no. 6, 1620–1649.
33. ———, *The gap between near commutativity and almost commutativity in symplectic category*, Electron. Res. Announc. Math. Sci. **20** (2013), 71–76.
34. L. Buhovsky, M. Entov, and L. Polterovich, *Poisson brackets and symplectic invariants*, Selecta Math. (N.S.) **18** (2012), no. 1, 89–157.
35. L. Buhovsky and Y. Ostrover, *On the uniqueness of Hofer's geometry*, Geom. Funct. Anal. **21** (2011), no. 6, 1296–1330.
36. D. Burago, S. Ivanov, and L. Polterovich, *Conjugation-invariant norms on groups of geometric origin*, Groups of Diffeomorphisms (R. Penner, D. Kotschick, T. Tsuboi, N. Kawazumi, T. Kitano, and Y. Mitsumatsu, eds.), Adv. Stud. Pure Math., vol. 52, Math. Soc. Japan, Tokyo, 2008, pp. 221–250.
37. P. Busch, M. Grabowski, and P. J. Lahti, *Operational quantum physics*, Lecture Notes in Phys. New Ser. m Monogr., vol. 31, Springer, Berlin, 1995.
38. P. Busch, T. Heinonen, and P. Lahti, *Noise and disturbance in quantum measurement*, Phys. Lett. A **320** (2004), no. 4, 261–270.
39. ———, *Heisenberg's uncertainty principle*, Phys. Rep. **452** (2007), no. 6, 155–176.
40. P. Busch, P. Lahti, and R. F. Werner, *Noise operators and measures of RMS error and disturbance in quantum mechanics*, available at `arXiv:1312.4393`.
41. E. Calabi, *On the group of automorphisms of a symplectic manifold*, Problems in Analysis (Princeton, NJ, 1969) (R. C. Gunning, ed.), Princeton Univ. Press, Princeton, NJ, 1970, pp. 1–26.
42. D. Calegari, *scl*, MSJ Memoirs, vol. 20, Math. Soc. Japan, Tokyo, 2009.
43. F. Cardin and S. Vazzoler, *Symplectic rigidity and weak commutativity*, J. Symplectic Geom. **11** (2013), no. 2, 163–166.
44. F. Cardin and C. Viterbo, *Commuting Hamiltonians and Hamilton-Jacobi multi-time equations*, Duke Math. J. **144** (2008), no. 2, 235–284.

45. J. F. Cariñena, F. Falceto, and M. F. Rañada, *A geometric approach to a generalized virial theorem*, J. Phys. A **45** (2012), no. 39, 395210, 19.
46. Yu. V. Chekanov, *Lagrangian tori in a symplectic vector space and global symplectomorphisms*, Math. Z. **223** (1996), no. 4, 547–559.
47. ———, *Invariant Finsler metrics on the space of Lagrangian embeddings*, Math. Z. **234** (2000), no. 3, 605–619.
48. C. C. Conley and E. Zehnder, *The Birkhoff–Lewis fixed point theorem and a conjecture of V. I. Arnold*, Invent. Math. **73** (1983), no. 1, 33–49.
49. ———, *Morse-type index theory for flows and periodic solutions for Hamiltonian equations*, Comm. Pure Appl. Math. **37** (1984), no. 2, 207–253.
50. A. Ya. Dubovicki, *On differentiable mappings of an n-dimensional cube into a k-dimensional cube* (Russian), Mat. Sbornik N.S. **32(74)** (1953), 443–464.
51. D. Eisenbud, U. Hirsch, and W. Neumann, *Transverse foliations of Seifert bundles and self-homeomorphism of the circle*, Comment. Math. Helv. **56** (1981), no. 4, 638–660.
52. Ya. M. Eliashberg, *A theorem on the structure of wave fronts and its application in symplectic topology* (Russian), Funktsional. Anal. i Prilozhen. **21** (1987), no. 3, 65–72, 96.
53. Ya. M. Eliashberg and L. Polterovich, *Bi-invariant metrics on the group of Hamiltonian diffeomorphisms*, Internat. J. Math. **4** (1993), no. 5, 727–738.
54. ———, *Local Lagrangian 2-knots are trivial*, Ann. of Math. (2) **144** (1996), no. 1, 61–76.
55. ———, *Symplectic quasi-states on the quadric surface and Lagrangian submanifolds*, available at `arXiv:1006.2501`.
56. M. Entov, *Commutator length of symplectomorphisms*, Comment. Math. Helv. **79** (2004), no. 1, 58–104.
57. ———, *Quasi-morphisms and quasi-states in symplectic topology*, Proceedings of the ICM (Seoul, 2014), to appear.
58. M. Entov and L. Polterovich, *Calabi quasimorphism and quantum homology*, Int. Math. Res. Not. **30** (2003), 1635–1676.
59. ———, *Quasi-states and symplectic intersections*, Comment. Math. Helv. **81** (2006), no. 1, 75–99.
60. ———, *Symplectic quasi-states and semi-simplicity of quantum homology*, Toric Topology (M. Harada, Y. Karshon, M. Masuda, and T. Panov, eds.), Contemp. Math., vol. 460, Amer. Math. Soc., Providence, RI, 2008, pp. 47–70.
61. ———, *C^0-rigidity of the double Poisson bracket*, Int. Math. Res. Not. IMRN **6** (2009), 1134–1158.
62. ———, *Lie quasi-states*, J. Lie Theory **19** (2009), no. 3, 613–637.
63. ———, *Rigid subsets of symplectic manifolds*, Compos. Math. **145** (2009), no. 3, 773–826.
64. ———, *C^0-rigidity of Poisson brackets*, Symplectic Topology and Measure Preserving Dynamical Systems (Snowbird, UT, 2007) (A. Fathi, Y.-G. Oh, and C. Viterbo, eds.), Contemp. Math., vol. 512, Amer. Math. Soc., Providence, RI, 2010, pp. 25–32.
65. M. Entov, L. Polterovich, and D. Rosen, *Poisson brackets, quasi-states and symplectic integrators*, Discrete Contin. Dyn. Syst. **28** (2010), no. 4, 1455–1468.
66. M. Entov, L. Polterovich, and F. Zapolsky, *Quasi-morphisms and the Poisson bracket*, Pure Appl. Math. Q. **3** (2007), no. 4, 1037–1055.
67. A. Floer, *Symplectic fixed points and holomorphic spheres*, Comm. Math. Phys. **120** (1989), no. 4, 575–611.
68. A. Floer and H. Hofer, *Coherent orientations for periodic orbit problems in symplectic geometry*, Math. Z. **212** (1993), no. 1, 13–38.

69. K. Fukaya, Y.-G. Oh, H. Ohta, and K. Ono, *Toric degeneration and nondisplaceable Lagrangian tori in $S^2 \times S^2$*, Int. Math. Res. Not. IMRN **13** (2012), 2942–2993.
70. ———, *Spectral invariants with bulk, quasimorphisms and Lagrangian Floer theory*, available at `arXiv:1105.5124`.
71. J.-M. Gambaudo and É. Ghys, *Commutators and diffeomorphisms of surfaces*, Ergodic Theory Dynam. Systems **24** (2004), no. 5, 1591–1617.
72. É. Ghys, *Knots and dynamics*, International Congress of Mathematicians. Vol. I (Madrid, 2006) (M. Sanz-Solé, J. Soria, J. L. Varona, and J. Verdera, eds.), Eur. Math. Soc., Zürich, 2007, pp. 247–277.
73. A. B. Givental, *Nonlinear generalization of the Maslov index*, Theory of Singularities and Its Applications (V. I. Arnold, ed.), Adv. Soviet Math., vol. 1, Amer. Math. Soc., Providence, RI, 1990, pp. 71–103.
74. ———, *The nonlinear Maslov index*, Geometry of Low-Dimensional Manifolds. 2 (Durham, 1989) (S. K. Donaldson and C. B. Thomas, eds.), London Math. Soc. Lecture Note Ser., vol. 151, Cambridge Univ. Press, Cambridge, 1990, pp. 35–43.
75. A. M. Gleason, *Measures on the closed subspaces of a Hilbert space*, J. Math. Mech. **6** (1957), 885–893.
76. M. Gromov, *Pseudoholomorphic curves in symplectic manifolds*, Invent. Math. **82** (1985), no. 2, 307–347.
77. ———, *Partial differential relations*, Ergeb. Math. Grenzgeb. (3), vol. 9, Springer, Berlin, 1986.
78. D. J. Grubb and T. LaBerge, *Additivity of quasi-measures*, Proc. Amer. Math. Soc. **126** (1998), no. 10, 3007–3012.
79. ———, *Spaces of quasi-measures*, Canad. Math. Bull. **42** (1999), no. 3, 291–297.
80. D. Gu and S. S. Rappaport, *Mobile user registration in cellular systems with overlapping location areas*, 1999 IEEE 49th Vehicular Technology Conference. Vol. 1 (Houston, TX, 1999), IEEE Press, Piscataway, NJ, pp. 802–806.
81. V. Guillemin, *Star products on compact pre-quantizable symplectic manifolds*, Lett. Math. Phys. **35** (1995), no. 1, 85–89.
82. V. Guillemin and S. Sternberg, *Convexity properties of the moment mapping*, Invent. Math. **67** (1982), no. 3, 491–513.
83. M. Hayashi, *Quantum information: An introduction*, Springer, Berlin, 2006.
84. M. W. Hirsch, *Differential topology*, Vol. 33, Springer, New York, 1976. Grad. Texts in Math.
85. H. Hofer, *On the topological properties of symplectic maps*, Proc. Roy. Soc. Edinburgh Sect. A **115** (1990), no. 1-2, 25–38.
86. ———, *Estimates for the energy of a symplectic map*, Comment. Math. Helv. **68** (1993), no. 1, 48–72.
87. M. Hutchings, *Recent progress on symplectic embedding problems in four dimensions*, Proc. Natl. Acad. Sci. USA **108** (2011), no. 20, 8093–8099.
88. ———, *Lecture notes on Morse homology (with an eye towards Floer theory and pseudo-holomorphic curves)*, available at `http://www.math.umass.edu/~sullivan/797SG/hutchings-morse.pdf`.
89. S. Ishikawa, *Uncertainty relations in simultaneous measurements for arbitrary observables*, Rep. Math. Phys. **29** (1991), no. 3, 257–273.
90. B. Janssens, *Unifying decoherence and the Heisenberg principle*, available at `arXiv:quant-ph/0606093`.
91. A. Jenová and S. Pulmannová, *Characterizations of commutative POV measures*, Found. Phys. **39** (2009), no. 6, 613–624.
92. M. Khanevsky, *Hofer's metric on the space of diameters*, J. Topol. Anal. **1** (2009), no. 4, 407–416.
93. N. Kornechuk, *Exact constants in approximation theory*, translated by K. Ivanov, Encyclopedia Math. Appl., vol. 38, Cambridge Univ. Press, Cambridge, 1991.

94. F. Lalonde, *Energy and capacities in symplectic topology*, Geometric Topology (Athens, GA, 1993) (W. H. Kazez, ed.), AMS/IP Stud. Adv. Math., vol. 2, Amer. Math. Soc., Providence, RI, 1997, pp. 328–374.
95. F. Lalonde and D. McDuff, *The geometry of symplectic energy*, Ann. of Math. (2) **141** (1995), no. 2, 349–371.
96. ______, *Hofer's* L^∞*-geometry: energy and stability of Hamiltonian flows.* I, Invent. Math. **122** (1995), no. 1, 1–33; II, 35–69.
97. N. P. Landsman, *Mathematical topics between classical and quantum mechanics*, Springer Monogr. Math., Springer, New York, 1998.
98. S. Lang, *Algebra*, 3rd ed., Grad. Texts in Math., vol. 211, Springer, New York, 2002.
99. X. Ma and G. Marinescu, *Toeplitz operators on symplectic manifolds*, J. Geom. Anal. **18** (2008), no. 2, 565–611.
100. J. Marsden and A. Weinstein, *Reduction of symplectic manifolds with symmetry*, Rep. Mathematical Phys. **5** (1974), no. 1, 121–130.
101. H. Martens and W. M. de Muynck, *Nonideal quantum measurements*, Found. Phys. **20** (1990), no. 3, 255–281.
102. S. Massar, *Uncertainty relations for positive-operator-valued measures*, Phys. Rev. A (3) **76** (2007), no. 4, 042114, 5.
103. D. McDuff, *Geometric variants of the Hofer norm*, J. Symplectic Geom. **1** (2002), no. 2, 197–252.
104. ______, *Hamiltonian* S^1*-manifolds are uniruled*, Duke Math. J. **146** (2009), no. 3, 449–507.
105. ______, *Loops in the Hamiltonian group: a survey*, Symplectic Topology and Measure Preserving Dynamical Systems (Snowbird, UT, 2007) (A. Fathi, Y.-G. Oh, and C. Viterbo, eds.), Contemp. Math., vol. 512, Amer. Math. Soc., Providence, RI, 2010, pp. 127–148.
106. ______, *Monodromy in Hamiltonian Floer theory*, Comment. Math. Helv. **85** (2010), no. 1, 95–133.
107. D. McDuff and D. Salamon, *Introduction to symplectic topology*, 2nd ed., Oxford Math. Monogr., Oxford Univ. Press, New York, 1998.
108. ______, *J-holomorphic curves and symplectic topology*, 2nd ed., Amer. Math. Soc. Colloq. Publ., vol. 52, American Mathematical Society, Providence, RI, 2012.
109. J. Milnor, *Morse theory*, Ann. of Math. Stud., vol. 51, Princeton Univ. Press, Princeton, NJ, 1963.
110. J. W. Milnor and J. D. Stasheff, *Characteristic classes*, Ann. of Math. Stud., vol. 76, Princeton Univ. Press, Princeton, NJ; University of Tokyo Press, Tokyo, 1974.
111. B. Misra, *Nonequilibrium entropy, Lyapounov variables, and ergodic properties of classical systems*, Proc. Nat. Acad. Sci. U.S.A. **75** (1978), no. 4, 1627–1631.
112. T. Miyadera and H. Imai, *Heisenberg's uncertainty principle for simultaneous measurement of positive-operator-valued measures*, Phys. Rev. A (3) **78** (2008), no. 5, 052119.
113. N. Monod, *An invitation to bounded cohomology*, International Congress of Mathematicians. Vol. II (Madrid, 2006) (M. Sanz-Solé, J. Soria, J. L. Varona, and J. Verdera, eds.), Eur. Math. Soc., Zürich, 2006, pp. 1183–1211.
114. J. Moser, *On the volume elements on a manifold*, Trans. Amer. Math. Soc. **120** (1965), 286–294.
115. R. Narasimhan, *Analysis on real and complex manifolds*, Adv. Stud. Pure Math., vol. 1, Masson, Paris; North-Holland, Amsterdam, 1968.
116. M. A. Neumark, *On a representation of additive operator set functions*, C. R. (Doklady) Acad. Sci. URSS (N.S.) **41** (1943), 359–361.
117. Y.-G. Oh, *Floer cohomology of Lagrangian intersections and pseudo-holomorphic disks.* II: $(\mathbb{C}P^n, \mathbb{R}P^n)$, Comm. Pure Appl. Math. **46** (1993), no. 7, 995–1012.

118. ———, *Chain level Floer theory and Hofer's geometry of the Hamiltonian diffeomorphism group*, Asian J. Math. **6** (2002), no. 4, 579–624.
119. ———, *Construction of spectral invariants of Hamiltonian paths on closed symplectic manifolds*, The Breadth of Symplectic and Poisson Geometry (J. E. Marsden and T. S. Ratiu, eds.), Progr. Math., vol. 232, Birkhäuser Boston, Boston, MA, 2005, pp. 525–570.
120. ———, *Lectures on Floer theory and spectral invariants of Hamiltonian flows*, Morse Theoretic Methods in Nonlinear Analysis and in Symplectic Topology (Montréal, QC, 2004) (P. Biran, O. Cornea, and F. Lalonde, eds.), NATO Sci. Ser. II Math. Phys. Chem., vol. 217, Springer, Dordrecht, 2006, pp. 321–416.
121. ———, *Symplectic topology and Floer homology*, available at `http://www.math.wisc.edu/~oh/all.pdf`.
122. K. Ono, *Floer–Novikov cohomology and the flux conjecture*, Geom. Funct. Anal. **16** (2006), no. 5, 981–1020.
123. Y. Ostrover, *A comparison of Hofer's metrics on Hamiltonian diffeomorphisms and Lagrangian submanifolds*, Commun. Contemp. Math. **5** (2003), no. 5, 803–811.
124. ———, *Calabi quasi-morphisms for some non-monotone symplectic manifolds*, Algebr. Geom. Topol. **6** (2006), 405–434 (electronic).
125. Y. Ostrover and I. Tyomkin, *On the quantum homology algebra of toric Fano manifolds*, Selecta Math. (N.S.) **15** (2009), no. 1, 121–149.
126. M. Ozawa, *Uncertainty relations for joint measurements of noncommuting observables*, Phys. Lett. A **320** (2004), no. 5-6, 367–374.
127. ———, *Universal uncertainty principle in the measurement operator formalism*, J. Opt. B Quantum Semiclass. Opt. **7** (2005), no. 12, S672–S681.
128. C. Pearcy and A. Shields, *Almost commuting matrices*, J. Funct. Anal. **33** (1979), no. 3, 332–338.
129. A. Peres, *Quantum theory: concepts and methods*, Fundam. Theor. Phys., vol. 57, Kluwer, Dordrecht, 1993.
130. S. Piunikhin, D. Salamon, and M. Schwarz, *Symplectic Floer–Donaldson theory and quantum cohomology*, Contact and Symplectic Geometry (Cambridge, 1994) (C. B. Thomas, ed.), Publ. Newton Inst., vol. 8, Cambridge Univ. Press, Cambridge, 1996, pp. 171–200.
131. L. Polterovich, *Symplectic displacement energy for Lagrangian submanifolds*, Ergodic Theory Dynam. Systems **13** (1993), no. 2, 357–367.
132. ———, *The geometry of the group of symplectic diffeomorphisms*, Lectures Math. ETH Zurich, Birkhäuser, Basel, 2001.
133. ———, *Quantum unsharpness and symplectic rigidity*, Lett. Math. Phys. **102** (2012), no. 3, 245–264.
134. ———, *Symplectic geometry of quantum noise*, Comm. Math. Phys. **327** (2014), no. 2, 481–519.
135. L. Polterovich and Z. Rudnick, *Stable mixing for cat maps and quasi-morphisms of the modular group*, Ergodic Theory Dynam. Systems **24** (2004), no. 2, 609–619.
136. L. Polterovich and K. F. Siburg, *On the asymptotic geometry of area-preserving maps*, Math. Res. Lett. **7** (2000), no. 2-3, 233–243.
137. P. Py, *Quasi-morphismes de Calabi et graphe de Reeb sur le tore*, C. R. Math. Acad. Sci. Paris **343** (2006), no. 5, 323–328.
138. ———, *Quasi-morphismes et invariant de Calabi*, Ann. Sci. École Norm. Sup. (4) **39** (2006), no. 1, 177–195.
139. J. Robbin and D. Salamon, *The Maslov index for paths*, Topology **32** (1993), no. 4, 827–844.
140. O. Rousseau, *Quelques propriétés du complexe de Morse-Novikov*, M.Sc. Thesis, Université de Montréal, 2005.

141. Yu. B. Rudyak and F. Schlenk, *Minimal atlases of closed symplectic manifolds*, Commun. Contemp. Math. **9** (2007), no. 6, 811–855.
142. M. Schlichenmaier, *Berezin–Toeplitz quantization for compact Kähler manifolds. A review of results*, Adv. Math. Phys. (2010), Art. ID 927280, 38.
143. M. Schwarz, *Morse homology*, Progr. Math., vol. 111, Birkhäuser, Basel, 1993.
144. ———, *On the action spectrum for closed symplectically aspherical manifolds*, Pacific J. Math. **193** (2000), no. 2, 419–461.
145. S. Seyfaddini, *C^0-limits of Hamiltonian paths and the Oh–Schwarz spectral invariants*, Int. Math. Res. Not. IMRN **21** (2013), 4920–4960.
146. D. Shakhmatov, unpublished.
147. E. Shelukhin, *The action homomorphism, quasimorphisms and moment maps on the space of compatible almost complex structures*, Comment. Math. Helv. **89** (2014), no. 1, 69–123.
148. A. I. Shtern, *The Kazhdan–Milman problem for semisimple compact Lie groups* (Russian), Uspekhi Mat. Nauk **62** (2007), no. 1(373), 123–190; English transl., Russian Math. Surveys **62** (2007), no. 1, 113–174.
149. J.-C. Sikorav, *Systèmes hamiltoniens et topologie symplectique*, Universitá di Pisa, ETS Editrice, Pisa, 1990.
150. S. Smale, *Diffeomorphisms of the 2-sphere*, Proc. Amer. Math. Soc. **10** (1959), 621–626.
151. W. Stulpe, S. Gudder, and J. Hagler, *An uncertainty relation for joint position-momentum measurements*, Found. Phys. Lett. **1** (1988), no. 3, 287–292.
152. T. Sunada, *Riemannian coverings and isospectral manifolds*, Ann. of Math. (2) **121** (1985), no. 1, 169–186.
153. B. Sz.-Nagy, *Extensions of linear transformations in Hilbert space which extend beyond this space*, translated by L. F. Boron, appendix to F. Riesz and B. Sz.-Nagy, Functional Analysis, Frederick Ungar Publishing Co., New York, 1960.
154. M. Usher, *Spectral numbers in Floer theories*, Compos. Math. **144** (2008), no. 6, 1581–1592.
155. ———, *Duality in filtered Floer–Novikov complexes*, J. Topol. Anal. **2** (2010), no. 2, 233–258.
156. ———, *Deformed Hamiltonian Floer theory, capacity estimates and Calabi quasimorphisms*, Geom. Topol. **15** (2011), no. 3, 1313–1417.
157. C. Viterbo, *Symplectic topology as the geometry of generating functions*, Math. Ann. **292** (1992), no. 4, 685–710.
158. C. Voisin, *Hodge theory and complex algebraic geometry.* I, translated by L. Schneps, Cambridge Stud. Adv. Math, vol. 76, Cambridge Univ. Press, Cambridge, 2002.
159. J. von Neumann, *Mathematical foundations of quantum mechanics*, translated by R. T. Beyer, Princeton Univ. Press, Princeton, NJ, 1955.
160. C. Wendl, *Lectures on holomorphic curves in symplectic and contact geometry*, available at `arXiv:1011.1690`.
161. R. F. Wheeler, *Quasi-measures and dimension theory*, Topology Appl. **66** (1995), no. 1, 75–92.
162. J. A. Yorke, *Periods of periodic solutions and the Lipschitz constant*, Proc. Amer. Math. Soc. **22** (1969), 509–512.
163. F. Zapolsky, *Quasi-states and the Poisson bracket on surfaces*, J. Mod. Dyn. **1** (2007), no. 3, 465–475.
164. ———, *On almost Poisson commutativity in dimension two*, Electron. Res. Announc. Math. Sci. **17** (2010), 155–160.
165. ———, *Isotopy-invariant topological measures on closed orientable surfaces of higher genus*, Math. Z. **270** (2012), no. 1-2, 133–143.

Notation Index

$a \cap b$	intersection product in homology, 156
$A(x)$	operator valued expectation, 134
$A * B$	quantum product, 175
$\mathcal{A}_F$	action functional, 51
$\hat{A}$	POVM associated to a von Neumann observable, 131
$\mathcal{B}(F)$	singly generated subalgebra of $C(X)$, 70
$B \to A$	smearing of POVMs, 139
$c(a, F)$	spectral numbers of smooth functions, 147
	spectral invariants in Floer theory, 179
$c(a, \phi)$	spectral invariants in Floer theory, 180
c_1	first Chern class, 16
Cal	Calabi homomorphism, 49
Cal_U	Calabi homomorphism on $\mathcal{G}(U)$, 57
$\mathrm{CF}(F)$	Floer chain complex, 169
$\mathrm{CF}_i(F)$	$\mathbb{Z}_2$-graded Floer chain complex, 169
$C_c^\infty(M)$	smooth compactly supported functions, 21
$\mathrm{CM}_*(F)$	Morse chain complex, 153
$C(M)$	space of continuous functions, 65
$\mathrm{CN}_*(\alpha)$	Morse–Novikov chain complex, 159
$\mathrm{CO}(X)$	subsets of X that are open or closed, 72
CovDim	covering dimension, 74
$\mathrm{Crit}(F)$	set of critical points of F, 151
$\mathrm{Crit}_k(F)$	set of critical points of Morse index k, 151
CZ	normalized Conley–Zehnder index, 44
$\overline{\mathrm{CZ}}$	Conley–Zehnder index, 44
Δ_A	noise operator, 134
$\mathrm{Diff}(M)$	group of compactly supported diffeomorphisms, 1
$\mathbb{E}(\hat{A}, \xi)$	expectation of a von Neumann observable at a pure state, 135
e_c	spectral displacement energy, 54
e_H	(Hofer's) displacement energy, 13
$\mathrm{End}(H)$	space of bounded linear operators on H, 134
$\mathcal{F}$	product function space, 119
$\mathcal{F}_4(X_0, X_1, Y_0, Y_1)$	auxiliary function space, 103
$\mathcal{F}_4'(X_0, X_1, Y_0, Y_1)$	auxiliary function space, 104
$\mathrm{Fix}(f)$	fixed point set of f, 5
$\mathcal{G}(U)$	image of $\widetilde{\mathrm{Ham}}(U) \to \widetilde{\mathrm{Ham}}(M)$, 55

GW_j	Gromov–Witten invariants, 173
$\mathbb{H}$	hyperbolic half-plane, 37
$\mathcal{H}$	space of normalized Hamiltonians, 6
	as the Lie algebra of Ham, 9
$\mathrm{Ham}(M,\omega)$	group of Hamiltonian diffeomorphisms, 6
	as a Lie group, 9
$\widetilde{\mathrm{Ham}}(M,\omega)$	universal cover of $\mathrm{Ham}(M,\omega)$, 13
$\mathrm{HF}_*(F)$	Floer homology of F, 170
$\mathrm{HF}_*(M)$	Floer homology of M, 170
$\mathrm{HM}_*(F,\rho)$	Morse homology of a generic pair, 154
$\mathrm{HM}_*(M)$	Morse homology of M, 155
ind	Morse index, 151
I_ω	linear isomorphism $E \to E^*$ induced by ω, 2
	bundle isomorphism $TM \to T^*M$ induced by ω, 3
$\mathcal{K}$	field of Laurent series, 159
κ_{cl}	classical measure of noncommutativity, 128
κ_{q}	quantum measure of noncommutativity, 130
$\mathcal{L}(H)$	space of Hermitian operators on a Hilbert space, 68
$\Lambda(\mathbb{R}^{2n})$ or Λ	Lagrangian Grassmannian, 40
ΛM	space of contractible loops in M, 52
$\mathcal{L}_X$	Lie derivative, 4
$\mathcal{M}(\tilde{z}_-,\tilde{z}_+)$	space of Floer connecting trajectories, 164
$\overline{\mathcal{M}}(\tilde{z}_-,\tilde{z}_+)$	moduli space of unparametrized Floer trajectories, 165
$\widetilde{\mathcal{M}}(\tilde{z}_-,\tilde{z}_+)$	compactified moduli space, 169
$\mathrm{Maslov}(\gamma)$	Maslov index of a loop in $\mathrm{Sp}(2n)$, 44
$\mathcal{M}_J(\alpha,\beta,\gamma;j)$	space of holomorphic spheres passing through 3 cycles, 173
μ	Maslov quasi-morphism on $\widetilde{\mathrm{Sp}}(2n)$, 43
	Maslov class of a Lagrangian submanifold, 95
μ_L	(non-homogeneous) Maslov quasi-morphism, 43
$\mathcal{N}(A)$	inherent noise of a POVM, 140
$\mathcal{N}(A,\xi)$	inherent noise of a POVM at a pure state, 139
N_J	Nijenhuis tensor, 15
$\Vert F\Vert$	uniform norm, 11
ν_c	spectral pseudo-norm, 53
$([\omega],A)$	symplectic area of $A \in \pi_2(M)$, 161
$[\omega]$	symplectic area class of a Lagrangian, 94
	symplectic area functional $\pi_2(M) \to \mathbb{R}$, 161
ω^*	symplectic form on the dual space, 30
ω_0	standard symplectic form on $\mathbb{R}^{2n}$, 3
Ω_N	set of integers $1,\dots,N$, 131
$\mathcal{P}(F)$	set of contractible periodic orbits, 164
$\widetilde{\mathcal{P}}(F)$	all lifts of contractible periodic orbits, 169
pb_4	Poisson bracket invariant of a quadruple, 103
$\mathrm{pb}(\mathcal{U})$	Poisson bracket invariant of a cover, 128
Φ_{PSS}	PSS map, 178
$\overline{\pi}_2(M)$	$\pi_2(M)/\ker[\omega]$, 162
$\Pi(F,G)$	measure of noncommutativity, 59

$\mathrm{PSL}(2,\mathbb{R})$	isometry group of the hyperbolic half-plane, 37
$\mathrm{QH}(M)$	quantum homology, 173
Q_N	unit cube $[0,1]^N$, 138
$Q(\Omega)$	measurable functions $\Omega \to [0,1]$, 138
$r(F)$	growth of one-parameter subgroup in Hofer's norm, 97
$r_{F,G}$	profile function, 119
r_L	Robbin–Salamon index, 41
rot	Poincare rotation number, 39
S	positive generator of $\overline{\pi}_2(\mathrm{M})$, 162
sgrad	symplectic gradient, 3
Σ_L	Maslov–Arnold cycle, 41
sign	signature of a quadratic form, 41
$\mathfrak{sp}(2n)$	Lie algebra of $\mathrm{Sp}(2n)$, 44
$\mathrm{Sp}(2n)$	symplectic linear group, 30
$\mathrm{Sp}(E,\omega)$	group of linear symplectomorphisms, 30
spec	action spectrum, 52
$\mathrm{Symp}(M,\omega)$	group of compactly supported symplectomorphisms, 1
$\mathbb{V}(\hat{A},\xi)$	variance of a von Neumann observable at a pure state, 135
$w(U)$	spectral width, 57
$W^{\mathrm{s}}(x)$	stable manifolds of a critical point, 151
$W^{\mathrm{u}}(x)$	unstable manifolds of a critical point, 151
$[x,y]$	commutator of group elements, 23

Subject Index

action
  filtration, 179
  functional, 51
  spectrum, 52
almost complex structure, 15
  compatible with a symplectic form, 16
  integrable, 15
approximate measurement, 136
  joint, 136
  unbiased, 136
Arnold conjecture, 5, 170
Atiyah–Guillemin–Sternberg theorem, 93

Barge-Ghys quasimorphism, 37
Berezin–Toeplitz quantization, 129
  and POVMs, 131
bubbling, 167
Buhovsky's 2/3 law, 25

C^0-rigidity
  of symplectomorphisms, 2, 29
  of the Poisson bracket, 21
Calabi
  homomorphism, 49
    on exact symplectic manifolds, 50
  property, 62
  quasi-morphism, 63
Cartan's formula, 4
characteristic exponent, 150
Chern class, *see also* first Chern class
C^k-topology, *see also* Whitney topology
class function, 36
Clifford torus
  in $\mathbb{C}P^2$, 86
  in $\mathbb{C}P^n$, 93
  in $S^2 \times \cdots \times S^2$, 90
complex projective space, *see also* $\mathbb{C}P^n$
complex structure, *see also* almost complex structure, integrable
Conley–Zehnder index, 44
contact type hypersurface, 82
correspondence principle, 129
covering dimension, 74
$\mathbb{C}P^n$, 18
  quantum homology, 175
cusp trajectory, *see also* Floer, trajectory, cusp

Darboux
  chart, 3
  Theorem, 3
displaceable set, 12
displacement energy
  Hofer's, 13
  spectral, 54

Eliashberg–Gromov rigidity theorem, 2, 29
energy conservation law, 17

Finsler metric, 10
  on Ham, 11
first Chern class, 16
fixed point
  nondegenerate, 5
Floer
  equation, 164
  homology
    and singular homology, 170
    isomorphic to quantum homology, 178
    of F, 170
    of M, 170
    ring structure, *see also* pair of pants product
  trajectory, 165
    broken, 167
    cusp, 168

flux conjecture, 8
Frobenius algebra, 176
Fubini-Study form, 18

Gromov–Witten invariants, 173
  dimension formula, 174
  with the fundamental class, 174
group of Hamiltonian diffeomorphisms, 6
  as a Lie group, 9
  is a group, 6

Hadamard–Kolmogorov–Landau inequality
  classical, 34
  symplectic version, 34
Hamilton equation, 3
  in Darboux coordinates, 4
Hamiltonian
  chord, 107
  diffeomorphism, 3
    preserves the symplectic form, 3
  flow, 3
  function, 3, 6
    autonomous, 6
    normalized, 6
    of composition, *see also* product formula
  torus action, 92
  vector field, *see also* symplectic, gradient
Hofer
  displacement energy, *see also* displacement energy, Hofer's
  metric, 11
    is nondegenerate, 11, 61
    lift to the universal cover, 14

integrable systems, 93
intersection product, 156

J-holomorphic curve, 17
John ellipsoid, 12

Kähler manifold, 16

Lagrangian
  Grassmannian, 40
  knots, 95
  submanifold, 93
    monotone, *see also* monotone, Lagrangian submanifold
  subspace, 40
least action principle, 52, 163

Liouville
  vector field, 82
Liouville's theorem, 4

Möbius transformations, 37
magnitude of noncommutativity
  classical, 128
  quantum, 130
Markov kernel, 138
Marsden–Weinstein reduction, 17
Maslov
  class
    and Lie quasi-states on $\mathfrak{sp}(2n)$, 87
    in $H^1(\Lambda, \mathbb{Z})$, 43
    of a Lagrangian submanifold, 95
  index
    of a loop of Lagrangians, 43
    of a loop of symplectic matrices, 44
  quasi-morphism
    on $\widetilde{\mathrm{SL}}(2,\mathbb{R})$, 39
    on $\widetilde{\mathrm{Sp}}(2n)$, 43
Maslov–Arnold cycle, 41
median quasi-state, 66
  as a reduced quasi-state, 86
  is a symplectic quasi-state, 81
minimal Chern number, 162
moment map, 89
monotone
  Lagrangian submanifold, 95
  symplectic manifold, 161
Morse
  chain complex, 153
  function, 5
  homology
    and singular homology, 155
    of (F, ρ), 154
    of M, 155
    product, 157
  index, 151
  inequalities, 5

Naimark dilation, 132
Newlander-Nirenberg theorem, 15
Newton's second law, 4
Nijenhuis tensor, 15
noise operator of a POVM, *see also* POVM, noise operator
non-Archimedean absolute value, 177
nondegenerate element of $\widetilde{\mathrm{Ham}}$, 52
nondisplaceable fiber theorem, 90
nonsqueezing theorem, 127

pair of pants product, 171

partial quasi-additivity, 60
partial symplectic quasi-state, *see also* quasi-state, symplectic, partial
pb_4 invariant, 103
  behavior under products, 105
  behavior under symplectic embeddings, 105
  semicontinuity, 105
perfect group, 39
Poincare
  duality for spectral invariants, 180
  rotation number, 39
Poisson
  bracket, 9
    inequality, 59, 62
    lower semicontinuity, 21
  commuting
    continuous functions, 21
    functions, 10
positive operator valued measure, *see also* POVM
POVM, 131
  inherent noise, 140
  inherent noise at a pure state, 139
  integration with respect to, 131
  noise operator, 134
  on a finite set, 131
  smearing, 138
    operator, 138
  value space, 131
product formula, 6
profile function, 119
projector valued measure, 131
pseudo-norm, 45
  bounded, 45
  discrete, 45
pseudoholomorphic curve, *see also* J-holomorphic curve
PSS isomohprism, 178

quantizable symplectic manifold, 129
quantum homology, 173
  Frobenius structure, 176
  isomorphic to Floer homology, 178
quantum product, 175
quasi-linearity, 65
  topological quasi-state, 70
quasi-measure, 72
  integration with respect to, 74
quasi-morphism, 35
  defect of, 35
  homogeneous, 35
  homogenization, 35
quasi-state
  Lie, 86
  quantum, 69
  symplectic, 65
    partial, 58
    reduction of, 82
  topological, 70
quasi-subadditivity, 59

Reeb graph, 66
registration procedure
  classical, 141
  quantum, 142
Riemann surface, 17
Robbin–Salamon index, 41
rotation number, *see also* Poincare, rotation number

signature of a quadratic form, 41
simple group, 11
singly generated subalgebra of $C(X)$, 70
spectral
  invariant
    subadditive, *see also* subadditive spectral invariant
  invariants, 179
  number, 147
    Poincare duality, 149
    subadditivity, 148
  pseudo-norm, 53
  width, 57
spectrality
  spectral invariants, 180
  subadditive spectral invariant, 53
stability
  partial symplectic quasi-state, 58
  spectral invariants, 180
  subadditive spectral invariant, 53
stability axiom
  Calabi quasi-morphism, 62
stable manifold, 151
stem, 89
subadditive spectral invariant, 53
  Class I, 61
  Class II, 62
superheavy set, 91
symplectic
  area class of a Lagrangian, 94
  basis, 31
  form, 1
  gradient, 3
  linear group, 30
  manifold, 1

exact, 50
monotone, *see also* monotone, symplectic manifold
quantizable, *see also* quantizable symplectic manifold
quasi-state, *see also* quasi-state, symplectic
vector space, 3
symplectomorphism, 1

Translation on the torus are not Hamiltonian, 6

uncertainty
jump, 136
principle, 135
undistorted one-parameter subgroups, 97
uniformization theorem, 15
unsharpness principle, 140
unstable manifold, 151

vanishing property, 58
von Neumann observable, 131

weakly robust, 33
Whitney topology, 1

Name Index

Aarnes, 69, 71, 72
Aharonov, xii
Akveld, 111
Albers, xii
Ali, 138, 141
Arnold, 1, 3, 5, 43, 94
Arthurs, 137
Atiyah, 93
Audin, 150, 161

Banyaga, 7, 11, 49, 63, 150
Barge, 37
Barvinok, 12
Bell, 69
Ben Simon, 37
Berberian, 131
Berezin, 129, 132
Biran, xii, 99, 127, 128
Bohm, 69
Bordemann, 129
Borman, xii, 37, 63, 82, 83
Borthwick, 129
Bott, 16
Brandenbursky, 37
Buhovsky, xi, xii, 11, 22, 25, 27, 103, 106, 115, 117, 119, 121, 123, 124
Burago, 45
Busch, xii, 131, 132, 134, 136–138

Calabi, 49
Calegari, xii, 36
Cardin, 21, 29, 33
Carinena, 121
Carmeli, 138, 141
Charette, xii, 30
Chekanov, 95, 99
Chern, 16
Conley, 5, 44

Damian, 150, 161
Darboux, 3
de Muynck, 139
Dickstein, xii
Donaldson, 117, 150
Dubovickiĭ, 80

Eisenbud, 39
Eliashberg, xi, xii, 1, 2, 21, 29, 63, 95, 96
Entov, xi, xii, 21, 25, 33, 34, 37, 45, 57–59, 62, 86, 91, 99, 103, 115, 117, 119, 121–123, 173, 175, 180

Falceto, 121
Floer, 5, 170
Fukaya, 62, 63, 97, 117

Gambaudo, 37
Ghys, 37
Ginzburg, 54
Givental, 37
Gleason, 69
Goodman, 137
Grabowski, 131, 132, 137, 138
Gromov, 1, 2, 16, 21, 29, 114, 127, 150
Grubb, 74, 77
Gu, 141
Gudder, 137
Guillemin, 93, 129

Hagler, 137
Hayashi, 137
Heinonen, 134, 136
Heinosaari, 138, 141
Heisenberg, 135
Hermann, 69
Hirsch, 1, 39
Hofer, xii, 1, 11–13, 26
Humilière, xii
Hurtubise, 150
Hutchings, 127, 150, 155

Iacob, xii
Ishikawa, 137
Ivanov, 45

Janssens, 134, 137
Jencova, 138, 141
John, 12

Kalai, xii
Kaloshin, xii
Karshon, xii
Kawasaki, xii
Kelly, 137
Khanevsky, 99
Kindler, xii
Kislev, xii
Korneichuk, 119

LaBerge, 74, 77
Lahti, 131, 132, 134, 136–138
Lalonde, xii, 1, 11, 22
Landsman, 132
Lang, 177

Ma, 129
Marinescu, 129
Marsden, 17
Martens, 139
Massar, 134, 135
McDuff, xi, xii, 1, 3, 7, 11, 13, 14, 16, 22, 26, 30, 49, 50, 62, 63, 82, 98, 125, 127, 161, 165, 174, 175, 178, 180
Meinrenken, 129
Milnor, 11, 16, 148, 150
Misra, 142
Monod, 36
Moser, 13, 113, 114

Naimark, 132
Neumann, 39
Neumark, *see also* Naimark
Novikov, 158

Oh, xi, xii, 13, 22, 49, 52, 56, 61–63, 97, 173, 179, 180
Ohta, 62, 63, 97
Ono, 8, 62, 63, 97
Ostrover, xii, 11, 55, 62, 63
Ozawa, 134, 137

Pearcy, 125
Peres, 135
Piunikhin, 173, 178
Poincare, 37
Polterovich, I., xii, 141
Polterovich, V., xii
Pulmannova, 138, 141
Py, 37, 63

Ranada, 121
Rappaport, 141
Robbin, 40, 41
Rousseau, 158
Rudnick, 38
Rudyak, 128

Salamon, xi, xii, 3, 7, 13, 16, 30, 40, 41, 49, 50, 82, 111, 125, 127, 161, 165, 173–175, 178, 180
Samvelyan, xii, 106
Schelukhin, xii
Schlenk, xii, 128
Schlichenmaier, 129, 132
Schwarz, 49, 55, 150, 173, 178, 179
Seiberg, 150
Seidel, xii
Seyfaddini, xii, 55
Shakhmatov, 74
Shelukhin, 37
Shields, 125
Shtern, 36
Sikorav, 98
Smale, 16
Smith, xii
Sodin, xii
Sperber, xii
Stasheff, 16
Sternberg, 93
Stulpe, 137
Sunada, 2
Sz.-Nagy, 132

Toigo, 138, 141
Tu, 16
Tyomkin, 62, 63

Uribe, 129
Usher, 45, 61, 62, 180

Vazzoler, 29
Viterbo, 21, 33, 49, 173, 179
Voisin, 130
von Neumann, 68

Weinberger, xii
Weinstein, 17
Wendl, 16
Werner, 137

Wheeler, 74
Wilkinson, xii
Witten, 150

Xue, xii, 28

Yorke, 5

Zapolsky, xi, xii, 21, 37, 59, 73, 81, 122
Zehnder, 5, 44